Neurobiology
of the Leech

Neurobiology of the Leech

EDITED BY
Kenneth J. Muller
Carnegie Institution of Washington

John G. Nicholls
Stanford University School of Medicine

Gunther S. Stent
University of California, Berkeley

Cold Spring Harbor Laboratory
1981

Neurobiology of the Leech

Copyright 1981 by Cold Spring Harbor Laboratory
All rights reserved
Cover and book design by Emily Harste
Cover photo provided by Brian Payton
Printed in the United States of America

Library of Congress Cataloging in Publication Data
Main entry under title:

Neurobiology of the leech.

   Bibliography: p.
   Includes indexes.
   1. Neurobiology. 2. Leeches—Physiology.
I. Muller, Kenneth J. II. Nicholls, John G.
III. Stent, Gunther Siegmund, 1924-
IV. Cold Spring Harbor Laboratory.
ISBN 978-0-879691-46-8 (hardcover : alk. paper) --
ISBN 978-1-936113-09-5 (pbk. : alk. paper)
QP356.N4823   595.1´4504188   81-68893
                       AACR2

10 9 8 7 6 5 4 3 2

Authorization to photocopy items for internal or personal use, or the internal or personal use of specific clients, is granted by CSHL Press provided that the appropriate fee is paid directly to the Copyright Clearance Center (CCC). Write or call CCC at 222 Rosewood Drive, Danvers, MA 01923 (978-750-8400) for information about fees and regulations. Prior to photocopying items for educational classroom use, contact CCC at the above address. Additional information on CCC can be obtained at CCC Online at http://www.copyright.com/.

All CSHL Press publications may be ordered directly from Cold Spring Harbor Laboratory Press, 500 Sunnyside Boulevard, Woodbury, New York 11797-2924. Phone: 1-800-843-4388 (Continental U.S. and Canada). All other locations (516) 422-4100. FAX: (516) 422-4097. E-mail: cshlpress@cshl.edu. For a complete catalog of all Cold Spring Harbor Laboratory Press publications visit our World Wide Web Site http://www.cshlpress.com/.

This book is dedicated to the memory of Stephen Kuffler, friend and colleague, who inspired and encouraged so much of the work reported here.

Acknowledgments

We thank James D. Watson for proposing the idea of this book to us; year by year, his unfailing support and enthusiasm for the leech neurobiology course has provided an opportunity for contact among investigators working on the leech, for exchanging ideas and jointly developing new techniques. We also thank William Udry who has helped greatly with administrative matters. Jan Jansen and Ann Stuart who taught in the leech course in earlier years, offered valuable suggestions and criticisms of the book during its preparation.

The editors and authors acknowledge also with thanks the editorial help, communication of unpublished results, and provision of illustrations by R.L. Calabrese, E. Elliott, J. Fernandez, W.O. Friesen, P. Fuchs, L. Henderson, A.L. Kramer, D. Kuffler, C. Lent, E. Macagno, A. Mason, E.L. Petersen, E. Phillips, D. Ready, S. Scott, D.K. Stuart, W. Thompson, J.C. Weeks, and S. Zackson. We are extremely grateful to Nancy Ford and Judy Cuddihy for their extraordinary efficiency and kindness in helping us to produce this book.

Contents

1	**Introduction** Kenneth J. Muller, John G. Nicholls, and Gunther S. Stent	3
2	**Leech Biology and Behavior** Roy T. Sawyer	7
3	**History of Medicinal Leeching and Early Medical References** Brian Payton	27
4	**Structure of the Leech Nervous System** Brian Payton	35
5	**Sensory Cells and Motor Neurons** Susanna E. Blackshaw	51
6	**Synapses and Synaptic Transmission** Kenneth J. Muller	79
7	**Neural Circuits Generating Rhythmic Movements** Gunther S. Stent and William B. Kristan, Jr.	113

8	**Neurotransmitter Chemistry** Bruce G. Wallace	147
9	**Development of the Nervous System** David A. Weisblat	173
10	**Regeneration and Plasticity** Kenneth J. Muller and John G. Nicholls	197

Appendixes

A	**Killing Single Cells** Itzchak Parnas	227
B	**Immunological Identification of Specific Neurons** Birgit Zipser, Susan Hockfield, and Ronald McKay	235
C	**The Nervous System of the Leech: A Laboratory Manual**	249
D	**An Atlas of Neurons in the Leech, *Hirudo medicinalis***	277

Bibliography, 289
Name Index, 309
Subject Index, 315

Neurobiology
of the Leech

1

Introduction

Kenneth J. Muller
John G. Nicholls
Gunther S. Stent

Since the days of ancient Greece and Rome, leeches have been applied by physicians to patients suffering from diverse diseases such as epilepsy, angina, tuberculosis, meningitis, and hemorrhoids. By the 19th century, use of the medicinal leech was so prevalent that it became almost extinct in Western Europe, forcing Napoleon to import about 6 million leeches from Hungary in 1 year to treat his soldiers. This mania for leeching had at least one lasting benefit for contemporary biology—the medicinal application of leeches stimulated basic research on their reproduction, development, and anatomy. Thus, in the late 19th century, founders of experimental embryology, such as Whitman, chose the leech to follow the fates of early embryonic cells. Similarly, the nervous system of the leech was extensively studied by a roster of distinguished anatomists, including Sanchez, Ramon y Cajal, Gaskell, Del Rio Hortega, Odurih, and Retzius, who showed that it consisted of a chain of stereotyped ganglia each containing about 400 neurons. Interest in the leech thereafter declined, only to be rekindled in 1960 when Stephen Kuffler and David Potter at Harvard Medical School first applied modern neurophysiological techniques to its nervous system.

While searching for a preparation suitable for studying the neuroglial cells (the satellites that surround and support neurons in vertebrates and invertebrates), it became clear to Kuffler and Potter that the leech was the animal par excellence for which a wealth of anatomical information about neuroglia was already available. They showed that one could readily record from neurons as well as glial cells in leech ganglia and study their physiological interactions. Their work not only provided the basis for studying the properties of neuroglial cells in the vertebrate brain, but, equally important in the context of this book, it also introduced the leech to modern neurobiology.

As the chapter headings of this book illustrate, the nervous system of the leech has since been used extensively for the study of problems ranging from development to regeneration of neural connections following injury and from the biophysics of individual nerve cells to coordinated rhythmical movements of the animal. Such studies have moved forward by the introduction and use of a combination of physiological, pharmacological, anatomical, biochemical, and immunological techniques.

The stimulus for writing this book was provided by James Watson, director of Cold Spring Harbor Laboratory. In 1972 he invited us to organize an intensive laboratory course on the leech designed for advanced graduate students, postdoctoral fellows, and established scientists wishing to learn the concepts and techniques of modern neurobiology at the cellular level. In view of the gratifying interest aroused by this course and the growing number of investigators now working on the leech, Watson offered to sponsor a leech meeting at Cold Spring Harbor in 1981; and he further suggested that this meeting might provide the incentive and opportunity for publishing a collaborative overview of leech neurobiology.

Accordingly, our aim has been to provide a coherent account of work on the nervous system of the leech. As a result this is not a symposium volume made up of a series of research reports dealing with the most recent advances of the last 2 or 3 years. Rather, each author was asked to write a chapter that would provide a narrative account of the experimental work on a particular topic, and explain the significance for general neurobiology. Although the chapters are interrelated and all make use of basic knowledge presented in chapter 3, each chapter is designed to stand on its own. We hope, therefore, that this book will interest not only investigators working on the leech, but also students, of neurobiology in general.

It is clear that many different animals offer distinct advantages for studying the nervous system. Among the invertebrates, the sea hare, various snails, insects, nematodes, and crustaceans have all provided more than enough information to fill many volumes on the nervous system. Why, then, the leech? One answer was provided in 1885 by

Friedrich Nietzsche in *Thus Spake Zarathustra* in an extraordinary chapter entitled "The Leech." Zarathustra stumbles on a man whose arm is dangling in a pool of water collecting leeches. In the ensuing conversation, Zarathustra asks the man who he is, and he answers that he is the "conscientious man of the spirit." He then goes on to say:

> "Better to know nothing than half-know many things! Better to be a fool on one's own account than a wise man at the approval of others! I — go to the root of things:
> —What matter if it be great or small, if it be swamp or sky, a handsbreadth of ground is enough for me: if only it be thoroughly firm ground!
> —A handsbreadth of ground, one can stand upon that. In a truly conscientious knowledge there is nothing great and nothing small."
> "So perhaps you are an expert on the leech?" asked Zarathustra. "And do you probe the leech down to its ultimate roots, conscientious man?"
> "Oh, Zarathustra" answered the man who was trodden on, "That would be a colossal task, how could I undertake it!
> But what I am master of and expert on is the leech's *brain* — that is my world!"

Perhaps the main appeal of the leech is the beauty of the ganglion as it appears under the microscope, with its 400 or so neurons so recognizable and so familiar from segment to segment, from specimen to specimen, from species to species. As one looks at these limited aggregates of cells laid out in an orderly pattern, one cannot but marvel at how they, on their own, being the brain of the creature, are responsible for all its movements, hesitations, avoidance, mating, feeding, and sensations. In addition to the aesthetic pleasure provided by the preparation, we hope to communicate the intellectual excitement of trying to solve the circuitry and logic of a finite, well-organized nervous system one cell at a time.

2
Leech Biology and Behavior

Roy T. Sawyer
51 Llanyrnewdd
Penclawdd, West Glamorgan
South Wales SA4 3JN United Kingdom

Leeches (class Hirudinea) are predominantly bloodsucking worms and are related to earthworms (class Oligochaeta). Their relationship is so close that together both classes comprise the superclass Clitellata, a major group of segmented worms allied with myriapods and insects (Manton 1972; Anderson 1973).

The Oligochaeta are relatively unspecialized, both anatomically and behaviorally. Their locomotion depends upon contractions of circular and longitudinal muscles whose antagonism is mediated by a fluid-filled coelom, which is compartmentalized in each segment by intersegmental septa. During locomotion, the body is anchored to the substrate by stiff protractile bristles, or setae (chaetae), from which the class derives its name. The number of segments is variable, owing to continual addition throughout the life of the individual of new segments from a posterior growth zone. The architecture of the nervous system of oligochaetes reflects their behavioral simplicity and consists of a relatively unorganized (nonmedullated) segmental arrangement of neurons along the ventral midline.

The Hirudinea are anatomically and behaviorally more advanced than the Oligochaeta. Their locomotion is primarily achieved by means of suckers located at both ends of the body, a locomotory mode that requires well-developed functional coordination between front and rear (Gray et al. 1938). The presence of a caudal sucker is developmentally incompatible with a posterior growth zone, which explains why leeches do not add new segments in postembryonic development. In fact, all leeches have the same number of body segments. Constancy of segment number is a major evolutionary advance in biological organization, primarily because it allows a higher degree of specialization of different body regions. Leeches share this feature more with insects than with oligochaetes. Another anatomical departure of leeches from the oligochaete body plan is the loss of compartmentalization of the coelom by reduction of intersegmental septa (present only in the embryo). Moreover, in leeches the coelom itself is reduced by proliferation of filler connective tissue. The coelomic and vascular systems have become so modified that many leech species can be said to possess a true hemocoel (Sawyer and Fitzgerald 1981).

Leeches display a wide repertoire of behaviors, which is reflected in the highly stereotyped organization of their central nervous system (CNS). Since the few accounts of the behavior of leeches (e.g., Herter 1932) are not generally known, this chapter attempts to provide a perspective of the role of behavior in the biology of leeches.

BEHAVIOR AND SYSTEMATICS

The taxonomic classification of the Hirudinea (Table 1) can be considered in terms of diversity of feeding habits among the approximately 650 species of leeches. In fact, division of the Clitellata into the two

TABLE 1

Simplified classification of the Hirudinea

Superclass Clitellata
  Class Oligochaeta
  Class Hirudinea
    Order Rhynchobdellida
      Family Piscicolidae
      Family Glossiphoniidae
    Order Pharyngobdellida
      Family Erpobdellidae
    Order Gnathobdellida
      Family Hirudinidae
      Family Haemadipsidae

classes, Oligochaeta and Hirudinea, reflects this taxonomic principle. Oligochaetes feed on detritus and other organic material present in soil, whereas leeches are carnivorous. This primary difference in diet is reflected secondarily by anatomical differences in mouthparts and gut. Moreover, behavioral differences as seen in relation to this difference in feeding habits between oligochaetes and leeches resemble those found in mammals, where Herbivora (e.g., giraffe) are behaviorally dull and the Carnivora (e.g., lion) are clever and aggressive.

The class Hirudinea is comprised of three orders, Gnathobdellida, Rhynchobdellida, and Pharyngobdellida. The first two orders feed by bloodsucking, but they go about it in fundamentally different ways. Gnathobdellids possess jaws armed with teeth, with which they bite the host. They protect against the clotting of ingested blood in the gut by secreting a nonenzymatic polypeptide, hirudin, which specifically inhibits the clotting enzyme, thrombin (Bagdy et al. 1976). In contrast, rhynchobdellids possess a proboscis, which they insert into the body of the host. Rhynchobdellids protect against the clotting of blood by a totally different biochemical mechanism. They do so by secreting an enzyme, hementin, which dissolves clots by specific fibrin degradation after they have formed (Budzynski et al. 1981). These fundamental differences between gnathobdellids and rhynchobdellids make it likely that both orders have independently exploited the bloodsucking lifestyle. The third order, Pharyngobdellida (related to Gnathobdellida), has adapted predation on small invertebrates as its mode of feeding. These nonbloodsucking species, e.g., *Erpobdella punctata*, swallow their prey whole (Sawyer 1970; Young and Ironmonger 1979) and, accordingly, lack jaws and teeth.

Gnathobdellida

The Gnathobdellida are comprised of two major families, the aquatic Hirudinidae and the terrestrial Haemadipsidae. On the basis of feeding habits, the Hirudinidae are further divided into two subgroups, monostichodonts and distichodonts. Monostichodonts have well-developed jaws armed with numerous small, sharp teeth suitable for making shallow cuts in skin. This subgroup is behaviorally agile, a reflection of its specialization for sucking vertebrate blood. Representative examples are *Hirudo medicinalis* and *Macrobdella decora*, both of which feed on mammals and on frogs. Distichodonts have poorly developed jaws armed with a few blunt teeth suitable for seizing and masticating food. This subgroup (of which *Haemopis marmorata* is a member) feeds on invertebrates and is behaviorally less agile than its bloodsucking relatives.

Rhynchobdellida

The Rhynchobdellida are divided into two major families, Piscicolidae and Glossiphoniidae. The Piscicolidae have successfully invaded the sea from the poles to the tropics, where they feed on fish unavailable to other leeches. This family has undergone an extensive adaptive radiation with behavioral specialization of its own (Sawyer and Hammond 1973; Daniels and Sawyer 1975; Sawyer et al. 1975). The Glossiphoniidae live exclusively in fresh water, where most of them feed on blood of vertebrate hosts. Some of them, however, feed only on invertebrates.

BEHAVIOR AND ECOLOGY

The most important factor that determines the numbers and kinds of leeches in a given habitat is the availability and diversity of food (Sawyer 1974). Maximum utilization is made of potential food supply by adopting species-specific strategies in feeding behavior that lead to dietary partitioning. For example, vertebrates, which constitute the richest source of food for bloodsucking leeches, are partitioned by glossiphoniid species in the following manner: (1) Fish. Only a few glossiphoniids, e.g., *Actinobdella inequiannulata* (Daniels 1975), feed on fish. This is more the domain of freshwater Piscicolidae. (2) Amphibians. A few glossiphoniid species, e.g., *Batracobdella picta*, feed exclusively on amphibians. Adult *B. picta* feed on mature amphibians that come to the water to breed in the spring, whereas the young of this leech feed on tadpoles (Sawyer 1972). (3) Reptiles. The largest number of bloodsucking glossiphoniid species, e.g., *Placobdella parasitica* and *P. ornata*, feed on reptiles, especially turtles. (4) Birds. An aberrant group of species, e.g., *Theromyzon rude*, feeds exclusively in the nasal passages of aquatic birds, especially ducks. (5) Mammals. Very few glossiphoniid species feed on mammals, but the behavior of those that do, e.g., *Haementeria ghilianii*, is exceptionally complex.

In general, bloodsucking leeches do not restrict themselves to particular host species but feed on most members of their preferred vertebrate class. However, if no members of the preferred vertebrate class are available, most leech species are capable of feeding opportunistically on members of certain other vertebrate classes. Thus, the European glossiphoniid *Hemiclepsis marginata* normally feeds on fish, but it will feed on tadpoles (though not on adult frogs) when given the opportunity. Similarly, *H. ghilianii* prefers mammals, but it will feed on reptiles (but neither on amphibians nor fish), on which it can complete its entire life cycle.

Leech species parasitic on different classes of vertebrates differ also in the position of the mouth in the oral sucker and in the nature of crop caeca, salivary glands, and specialized organs (mycetomes) in the gut for harboring symbiotic bacteria. Indeed, these are the very characteristics used to diagnose glossiphoniid genera (Sawyer and Shelley 1976; Klemm 1981). In fact, the classification of bloodsucking Glossiphoniidae closely parallels that of their vertebrate hosts (Fig. 1), ranging from the generalized fish parasites to the advanced mammalian parasites. The idea that evolution of bloodsucking leeches has been moving toward parasitism of mammals has a sound biological explanation. During the evolution of vertebrates, the composition of their blood has changed from the nutritionally sparse fish blood to the erythrocyte-packed blood of mammals (Andrew 1965), which is a rich source of food for any bloodsucking animal (including the vampire bat). Thus, it is not by accident that the acme of glossiphoniid evolution, *H. ghilianii*, is by far the largest and most advanced species in the family. The concept of orthogenesis (i.e., evolution in a defined direction) in leech behavior has been proposed with respect to the evolution of brooding behavior within the Glossiphoniidae, culminating in *Marsupiobdella africana*, which broods its young in an internal pouch (Sawyer 1971).

In most aquatic habitats, availability of vertebrate blood is limited or sporadic, so that many leech species have successfully adapted to feeding on other sources of food, primarily aquatic invertebrates. Even in these nonbloodsucking species, between closely related species divergence in dietary specialization is the rule. For example, two species of the glossiphoniid genus *Helobdella* commonly live together in the same aquatic habitats in North America. One of these, *H. triserialis*, feeds almost exclusively on the body fluids of snails (Klemm 1975), whereas the other, *H. stagnalis*, feeds on the body fluids of oligochaetes, insect larvae, and crustaceans but rarely on snails (Davies et al. 1979). Interestingly, the diet of *H. stagnalis* changes quantitatively as the animal matures (Wrona 1981).

Similarly, it is not uncommon to find two or more species of the predaceous hirudinid genus *Haemopis* in the same habitat. Such species have similar but quantitatively different diets. This is, in part, a reflection of subtle differences in their foraging habits. For example, occasionally one can observe two common species of *Haemopis* feeding in proximity. Individuals of *H. marmorata* feed at, or slightly above, the edge of the water on snails and oligochaetes, whereas individuals of the much larger, but toothless, *H. grandis* feed at the bottom on various small invertebrates, most remarkably, on other leeches. A third species, *H. terrestris*, does not compete for food with the other

FIGURE 1

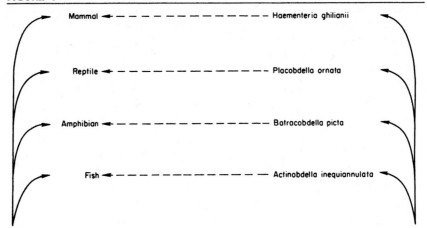

▲ Phylogenetic scheme showing the parallel systematic relationship of bloodsucking Glossiphoniidae (*right*) and their respective vertebrate hosts (*left*).

two and, instead, lives and feeds on earthworms on land (Sawyer 1981a). It is noteworthy that young *H. marmorata* eat only the soft parts of snails, leaving the shells, whereas larger individuals eat the entire snail, shell and all (Sawyer 1972).

The various species of leeches in a given habitat thus represent a rich diversity of behavioral strategies designed to obtain different kinds of food, both vertebrate and invertebrate. In light of this multiplicity of species-specific behaviors, how is one to reconcile the finding (Kramer and Goldman 1981) that the CNS of the glossiphoniid *H. ghilianii* is remarkably similar to that of the hirudinid *H. medicinalis*, a species so phylogenetically remote that it belongs to a different order? The probable explanation for this neurological conservatism lies in the observation that the elements of the behaviors displayed by leeches in foraging are qualitatively invariant: All leeches orient and move toward local disturbances in much the same basic manner. Diverse species specificities in host preference depend not on divergences in basic motor routines, but on contact recognition of the appropriate host, probably mediated by sensory mechanisms in the head.

BEHAVIOR AND THE INDIVIDUAL

How does a leech go about searching for a blood meal? This question can best be answered by detailed observation of the foraging habits of one representative bloodsucking leech in as natural a setting as possible. The giant glossiphoniid *H. ghilianii* is especially propitious for this purpose because it displays unusually marked changes in foraging behavior as it matures, a possible consequence of its extremely large adult size. Observations based on studies of the ecology of *H. ghilianii* in its native habitat (Sawyer 1978) revealed that, with respect to orienting and locomoting toward potential food, the foraging repertoire of the leech consists of only a small number of elementary behaviors. Furthermore, the growth and feeding history of an individual have profound influences on its behavior.

Growth and Feeding

The growth of *H. ghilianii* has been examined by Sawyer et al. (1981). This species is generally representative of bloodsucking leeches in that individuals reproduce after a small number of blood meals, in this case usually four. The interval between subsequent meals becomes progressively longer. Each blood meal results in an abrupt, approximately fivefold increase in body weight. The total weight of the animal remains constant between successive feedings, owing to the conversion of digested blood into body tissues. Consequently, growth in bloodsucking leeches consists of a series of discrete steps, or stages (Fig. 2). Each stage is associated with gross physiological changes, e.g., maturation of the male reproductive system occurs after the second feeding. As discussed further below, each stage of feeding is also associated with characteristic behavioral patterns.

In *H. ghilianii* and other bloodsucking leeches, feeding, or engorgement of blood, leads to an abrupt depression of behavioral activity. Using swimming as a criterion of activity, we examined the effect of feeding on the behavior of adult *H. ghilianii*. For this purpose, the duration of swimming episodes of leeches that were dropped into a cylinder of water of constant shape and volume was recorded as a function of time elapsed since the last blood meal. The result (Fig. 3) was that the animal will not swim immediately following a blood meal. (This postfeeding suppression of swimming is less evident in young *H. ghilianii*.) Over the next few days or weeks following feeding, the tendency to swim increases, until a plateau is reached, about the time the animal is ready to feed again. Thus, readiness to swim may be regarded as an operational definition of motivation or "hunger" of the leech.

Superimposed on the effect of feeding on readiness to swim is an important developmental factor: With each subsequent feeding, the tendency of *H. ghilianii* to swim is progressively diminished, so that swimming hardly recurs in *H. ghilianii* after the fourth feeding.

FIGURE 2

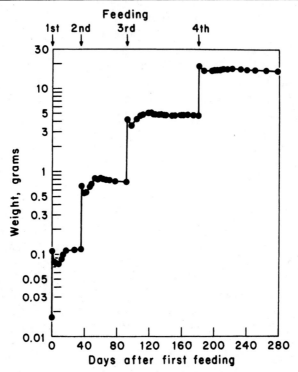

▲ Saltatory growth in an individual *H. ghilianii* from juvenile to sexually mature adult. The loss of weight in the first few days after feeding is attributable to the elimination of urine. (Reprinted, with permission, from Sawyer et el. 1981.)

Elementary Behaviors of the Foraging Repertoire

The foraging behavior of an unrestrained leech in seminatural surroundings is comprised of discrete elementary behaviors that lend themselves to quantitative analysis. Thanks to this procedure, one can observe quantitative changes in the relative frequency of expression of certain elementary behaviors as the animal matures (Table 2). Furthermore, one may determine whether a given elementary behavior is expressed autonomously or in association with other elementary behaviors.

Resting Postures. Leeches spend most of their time in resting postures, some of which are assumed by hungry leeches as the initial step

FIGURE 3

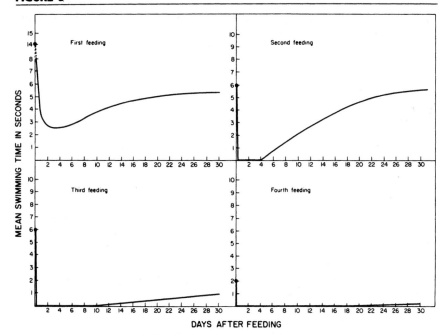

▲ Relationship in *H. ghilianii* of swimming to feeding following four consecutive blood meals. The curves are based on points representing the mean swimming times of 20–22 animals gently released in the water near the surface in a cylinder (38.5 cm in diameter filled with water to a depth of 22 cm; temperature 23–24°C). Note the long mean swimming times (about 14 sec) of individuals tested before their first meal, (cf. adult *H. medicinalis*). With each meal, swimming is depressed for varying lengths of time.

in foraging; such postures increase the likelihood of coming into contact with potential food. Leeches ready to forage are very responsive to light and mechanical stimuli. They tend to change location relatively frequently and often move near the water surface. Successful foraging results in individuals satiated with a blood meal. Such individuals, characterized by an overriding avoidance of light, seek out dark areas, such as the underside of a rock or submerged log. Satiated leeches remain quiescent for very long periods and are much less responsive to stimuli. This extreme quiescent, light-avoiding state minimizes contact with potential predators and allows undisturbed digestion of the ingested blood.

TABLE 2

Elementary behaviors of *H. ghilianii* whose frequency of expression in foraging changes progressively during maturation and growth

	Feeding stage (%)[a]		
	first (N=224)	second (N=485)	third (N=231)
Behaviors that decrease in frequency			
Body waving	26	6	0
Alert posture	28	22	8
Inchworm crawling	17	0.4	1
Swimming episode	19	6	0.4
Behaviors that increase in frequency			
Head movement	8	18	30
Slow shortening	1	7	9
Vermiform crawling	1	42	51

[a] Values are given as percentage of total elementary behaviors expressed by an individual of each respective stage (N=total number of elementary behaviors expressed).

Exploratory Behaviors. Virtually all leeches display movements reminiscent of exploration or searching. The biological significance of this exploratory behavior can only be presumed, but there is little doubt that such movements increase input of sensory information concerning the animal's immediate environment, as well as orient it toward the source of a local disturbance. Exploratory behavior is most readily and aggressively expressed by hungry animals in response to a variety of stimuli, especially water disturbance and changes in light intensity.

Two types of exploratory behaviors can be distinguished — head movement and body waving. In head movement (Fig. 4A), the anterior region of the body, primarily the head, bends irregularly and slowly. Head movement is characteristic of large leeches or of small leeches that are only moderately hungry. In body waving (Fig. 4B), the entire body pivots rapidly about the firmly attached caudal sucker. Body waving is characteristic of small leeches that are very hungry. The body-waving mode of exploration is replaced by the head-movement mode as the animal matures (Table 2).

Alert Posture. In response to various stimuli, a leech may extend itself to full length and remain motionless (Fig. 4C), occasionally for long

FIGURE 4

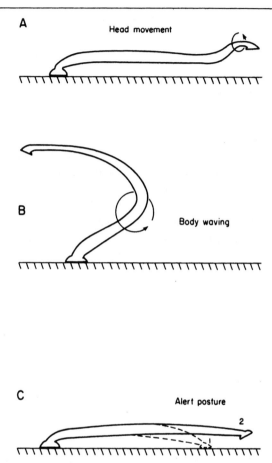

▲ (A) Head movement. (B) Body waving. (C) Alert Posture (2) following disturbance of an individual in resting posture (1). Note the rigid, linear configuration of the extended body.

periods. This alert posture can be presumed to reflect a state of higher excitation, in primed readiness for further sensory input. One may further speculate that the extended position of the body maximizes the ability of sensory structures in the skin to receive and/or orient toward the source of exogenous stimuli.

The supposition that alert posture reflects a primed excitatory state receives some quantitative support from the observed foraging repertoire of *H. ghilianii:* Alert posture is the behavior most frequently expressed prior to swimming both in the first-fed stage (comprising 34% of all behaviors preceding swimming) and in the second-fed stage (59%).

Shortening. Shortening of the body in response to external stimuli is behavior common to all leeches, as well as to oligochaetes. The biological significance of shortening in leeches has not been demonstrated, but at least sometimes it is defensive. Shortening may be slow or fast. In most instances, shortening is slow and mainly involves the anterior portion of the body. Relaxation is rapid. However, in response to sudden intense mechanical stimulation, especially to the head, the shortening response is usually fast and involves virtually the whole body, which can contract by as much as 70%. Here relaxation is slow.

The neuronal basis of shortening in leeches is partially understood and can be evoked by mechanosensory input to the longitudinal (L) motor neuron (see Chapters 5 and 6).

Crawling. Crawling is the only mode of locomotion characteristic of all Hirudinea. Two basic modes of crawling can be distinguished — inchworm crawling and vermiform crawling (Fig. 5). Both inchworm and vermiform crawling share body elongation as the initial phase of the crawling sequence. They differ in subsequent movements of the caudal end of the body and in the attachment position of the caudal sucker.

In vermiform crawling, the leech pulls up its caudal end without executing a very pronounced loop (Fig. 5 D', E'). During this movement, the annuli are sometimes erected, presumably affording the animal better traction against the substrate (reminiscent of setae erection in oligochaetes). Vermiform crawling would appear to be more effective than inchworm crawling on an unstable substrate, such as sand or mud. The resemblance of vermiform crawling to oligochaete locomotion is probably not coincidental.

In inchworm crawling, the leech pulls up the caudal end by bending the whole body into a distinct loop (Fig. 5D, E). The caudal sucker is attached close to and virtually touching the oral sucker, which is quickly released. Inchworm crawling would appear to be more efficient than vermiform crawling on a solid (smooth) substrate, such as a rock or plant, especially for small leeches.

Some leech species express only one or the other mode of crawling, but many species are capable of expressing either mode. In *H. ghilianii*, inchworm crawling is replaced by vermiform crawling as the animal matures (Tables 2 and 3).

FIGURE 5

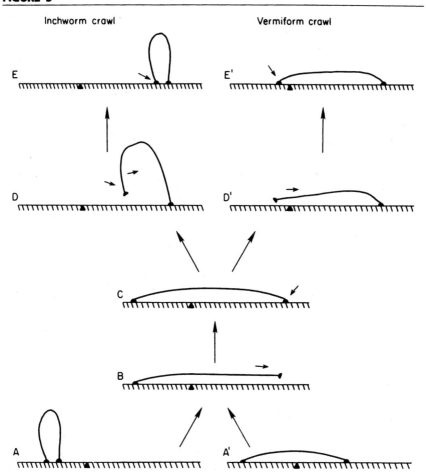

▲ Crawling behaviors in the Hirudinea. Schematic comparison of the major components of inchworm crawling (*left*) and vermiform crawling (*right*). Both behaviors begin (*A, A'*) with extension of the anterior body (*B*) and attachment of the oral sucker to the substrate (*C*). In inchworm crawling (*D, E*) the animal forms a midbody loop and positions the caudal sucker near the oral sucker. In vermiform crawling (*D', E'*), the posterior part of the body is pulled up a short distance from behind.

Swimming. Many leeches are capable of sustained swimming, the neuronal basis of which has been elucidated (Stent et al. 1978). Swimming serves primarily to transport the animal to a suitable source of

TABLE 3

Ontogenetic comparison of the relative frequencies of the locomotory behaviors expressed in the foraging repertoires by three successive postjuvenile stages of *H. ghilianii*

	Feeding stage (%)[a]					
	first		second		third	
Inchworm crawling	45	(37)	1	(2)	1	(1)
Vermiform crawling	4	(3)	87	(202)	98	(118)
Swimming episode	51	(41)	12	(29)	1	(1)
Total		(81)		(233)		(120)

[a] Values are percentages of total locomotory behaviors expressed by an individual of each stage in getting to a host (total number of elementary behaviors expressed [*N*] in parentheses).

food, especially under ecological conditions where crawling on the substrate is severely limited, e.g., on mud and sand, or where distances to be covered are great. Swimming can be initiated in response to water movement (Friesen and Dedwylder 1978).

Not all leech species are capable of swimming, and some species swim only under special circumstances.

Foraging Repertoire of *H. ghilianii*

Quantitative observations were made under seminatural conditions on the foraging of hungry *H. ghilianii* at different stages of maturation, using a turtle as a potential host. A typical foraging sequence displayed by an immature, first-fed-stage individual is summarized as follows (Fig. 6). When undisturbed for long periods, the leech either lies motionless (resting posture) among the surface vegetation or burrows in the muddy substrate. In response to an approaching host, the leech displays body waving for a few seconds, followed by alert posture. With further movements of the host, the leech initiates inchworm crawling, sometimes several episodes in succession, followed by body waving and alert posture. Upon further disturbances originating from movements of the host, the leech begins swimming, usually in short, irregular spirals. Swimming is terminated by attachment to a piece of floating vegetation. The locomotory sequences, inchworm crawling and swimming, are repeated until the leech is close to the host. Finally, and usually following swimming, the leech touches the host and immediately attaches itself.

Foraging at the adult third-fed stage is very different, and typically

FIGURE 6

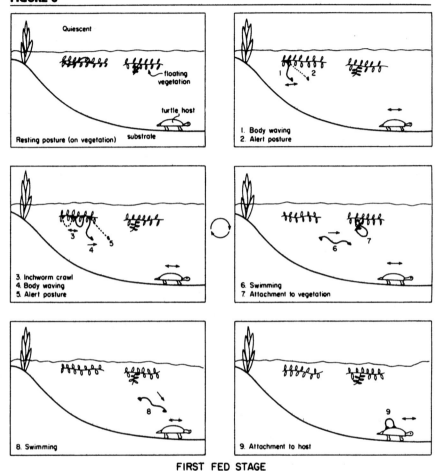

FIRST FED STAGE

▲ Typical sequence of elementary behaviors in the foraging repertoire displayed by first-fed individuals of *H. ghilianii* under seminatural conditions.

takes the following course (Fig. 7). The undisturbed leech lies motionless (resting posture), burrowed in the muddy substrate or hidden under an object. It responds to an approaching host by head movement, followed by an often prolonged sequence of vermiform crawlings. This is followed by head movement and yet another sequence of vermiform crawlings, usually oriented more precisely toward the sporadically active host. The simple sequence outlined above is repeated, interspersed occasionally with alert postures. The sequences of vermi-

FIGURE 7

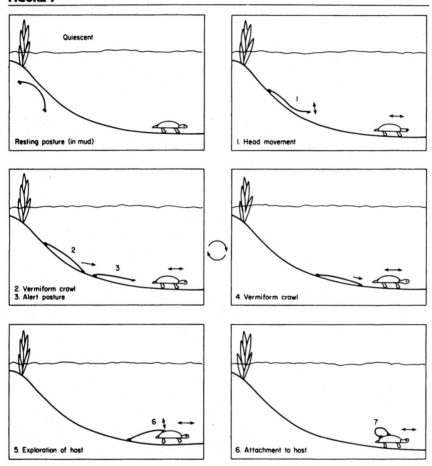

THIRD FED STAGE

▲ Typical sequence of elementary behaviors in the foraging repertoire displayed by third-fed individuals of *H. ghilianii* under seminatural conditions.

form crawling, attended by occasional changes in direction, maneuver the leech into proximity to the host, to which it may or may not attach.

Elementary behaviors that decrease in relative frequency during postembryonic development of *H. ghilianii* are swimming, inchworm crawling, alert posture, and body waving. Those that increase in relative frequency are vermiform crawling, slow shortening, and head movement (Table 2).

PHYLOGENY AND ONTOGENY OF BEHAVIOR

Observations on adult *H. medicinalis* under seminatural conditions similar to those already described indicate that this hirudinid species uses the same elementary behaviors for foraging as the glossiphoniid *H. ghilianii*. But the two species differ in the relative frequency with which they express a given elementary behavior (Table 4). One striking difference is the much higher tendency of adult *H. medicinalis* to swim. However, this interspecific difference pertains only to fully mature adults at the same level of "hunger;" juvenile (prefed) and first-fed *H. ghilianii* swim as readily as do adult *H. medicinalis*.

Observations on the terrestrial species *Haemadipsa zeylanica* (ground leech) and *H. picta* (bush leech) in a natural setting (Sawyer 1981b) show that these phylogenetically distant and ecologically peculiar species also use the standard elementary behaviors, except that their adults only rarely resort to vermiform crawling and seem incapable of swimming (Table 4).

Progressive decrease in frequency of swimming during postembryonic development of *H. ghilianii*, to the point of being virtually absent in mature adults (Table 3), is also encountered among other

TABLE 4

Relative frequencies of elementary locomotory behaviors expressed in the foraging repertoires of adults of four leech species

	Frequencies (%)[a]			
	Haementeria ghilianii	*Hirudo medicinalis*	*Haemadipsa zeylanica*	*Haemadipsa picta*
Inchworm crawling	1 (1)	3 (4)	97 (164)	98 (122)
Vermiform crawling	98 (118)	44 (60)	3 (5)	2 (2)
Swimming episode	1 (1)	53 (71)	0	0
Total	(120)	(135)	(169)	(124)

Observations on the aquatic species *H. ghilianii* and *H. medicinalis* were made under seminatural conditions constructed in the laboratory. Observations on the terrestrial species *H. zeylanica* and *H. picta* were made in nature.

[a] Values are percentages of total locomotory behaviors expressed by an individual of each species in getting to a host (total number of elementary behaviors expressed [N] in parentheses).

FIGURE 8

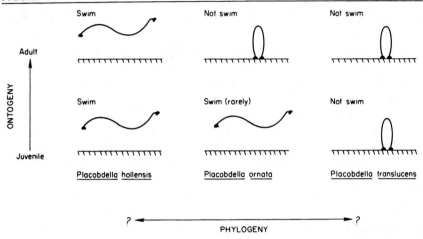

▲ Scheme showing a relationship between ontogenetic and phylogenetic development of swimming in American species of *Placbdella*. The direction of phylogenetic development can only be presumed but is not crucial to the argument.

glossiphoniid species. Juveniles of *P. ornata* and *P. parasitica* seeking their first blood meal are capable of sustained swimming (Sawyer 1972), but at all later stages these species seem incapable of swimming.

On the basis of swimming behavior, we can thus consider the following phylogenetic series of glossiphoniid species: (1) those that never swim at any stage in their life, e.g., *A. inequiannulata*, *B. picta*, and *Placobdella translucens*; (2) those that swim only as juveniles, e.g., *P. ornata* and *P. parasitica*; (3) those that swim primarily in younger stages, e.g., *H. ghilianii*; and (4) those that apparently swim well at virtually all stages in their lives, e.g., *Placobdella hollensis* (Sawyer 1972). This phylogenetic distribution of swimming among glossiphoniids suggests a relationship between phylogeny and ontogeny of this behavior (Fig. 8).

REFERENCES

Anderson, D.T. 1973. *Embryology and phylogeny in annelids and arthropods.* Pergamon Press, Oxford.

Andrew, W. 1965. *Comparative haematology.* Grune and Stratton, New York.

Bagdy, D., E. Barabas, L. Graf, T.E. Petersen, and S. Magnusson. 1976. Hirudin. *Methods Enzymol.* **45**:669–678.

Budzynski, A.Z., S.A. Olexa, B.S. Brizuela, R.T. Sawyer, and G.S. Stent. 1981. Anticoagulant and fibrinolytic properties of salivary proteins from the leech *Haementeria ghilianii*. *Proc. Soc. Exp. Biol. Med.* (in press).

Daniels, B.A. 1975. "On the biology of *Actinobdella inequiannulata* (Hirudinea, Glossiphoniidae) parasitic on *Catostomus* spp. in Algonquin Park." Masters thesis, University of Toronto, Ontario.

Daniels, B.A. and R.T. Sawyer. 1975. The biology of the leech *Myzobdella lugubris* infesting blue crabs and catfish. *Biol. Bull.* **148**:193–198.

Davies, R.W., F.J. Wrona, and F. Linton. 1979. A serological study of prey selection by *Helobdella stagnalis* (Hirudinoidea). *J. Animal Ecol.* **48**:181–194.

Friesen, W.O. and R. D. Dedwylder. 1978. Detection of low amplitude water movement: A new sensory modality in the medicinal leech. *Neurosci. Abstr.* **4**:380.

Gray, J., H.W. Lissman, and R.J. Pumphrey. 1938. The mechanism of locomotion in the leech *(Hirudo medicinalis)*. *J. Exp. Biol.* **15**:408–430.

Herter, K. 1932. Hirudinea, Egel. In *Biologie der Tiere Deutschlands* (ed. P. Schulze), vol. 12b, pp. 1–158. Berlin.

Klemm, D.J. 1975. Studies on the feeding relationships of leeches (Annelida: Hirudinea) as natural associates of mollusks. *Sterkiana* **58**:1–50 and **59**:1–20.

_____. 1981. Freshwater leeches (Annelida: Hirudinea) of North America. In *Biota of freshwater ecosystems identification manual.* Superintendent of Documents, U.S. Government Printing Office, Washington, D.C.

Kramer, A.P. and J.R. Goldman. 1981. The nervous system of the glossiphoniid leech *Haementeria ghilianii*. I. Identification of neurons. *J. Comp. Physiol.* (in press).

Manton, S.M. 1972. The evolution of arthropodal locomotory mechanisms, Part 10. *J. Linn. Soc. Lond. Zool.* **51**:203–400.

Sawyer, R.T. 1970. Observations on the natural history and behavior of *Erpobdella punctata* (Leidy) (Annelida: Hirudinea). *Am. Mid. Nat.* **83**:65–80.

_____. 1971. The phylogenetic development of brooding behavior in the Hirudinea. *Hydrobiologia* **37**:197–204.

_____. 1972. *North American freshwater leeches, exclusive of the Piscicolidae, with a key to all species.* University of Illinois Press, Urbana.

_____. 1974. Ecology of freshwater leeches. In *Pollution ecology of freshwater organisms* (ed. C.W. Hart and S.L.H. Fuller), pp. 81–142. Academic Press, New York.

_____. 1978. Domestication of the world's largest leech for developmental neurobiology. *Year Book of the American Philosophical Society*, pp. 212–213.

_____. 1981a. Terrestrial Leeches. In *Soil biology guide* (ed. D.L. Dindal). Interscience, New York.

_____. 1981b. An expedition to Borneo to study the aggressive behavior of land leeches, with collateral analysis of their anti-coagulants for medicinal purposes. *Year Book of the American Philosophical Society*. (In press.)

Sawyer, R.T. and S. Fitzgerald. 1981. Leech circulatory system. In *Invertebrate blood cells.* (ed. N.A. Radcliffe and A.F. Rowley), vol. 1, pp. 141–159.

Academic Press, London.

Sawyer, R.T. and D.L. Hammond. 1973. Distribution, ecology and behavior of the marine leech *Calliobdella carolinensis* (Annelida: Hirudinea), parasitic on the Atlantic menhaden in epizootic proportions. *Biol. Bull.* **143**:373-388.

Sawyer, R.T. and R.H. Shelley. 1976. New records and species of leeches (Annelida: Hirudinea) from North and South Carolina. *J. Nat. Hist.* **10**:65-97.

Sawyer, R.T., A.R. Lawler, and R.H. Overstreet. 1975. The marine leeches of the eastern United States and the Gulf of Mexico with a key to the species. *J. Nat. Hist.* **9**:633-667.

Sawyer, R.T., F. LePont, D.K. Stuart, and A.P. Kramer. 1981. Growth and reproduction of the giant glossiphoniid leech *Haementeria ghilianii*. *Biol. Bull.* **160**:322-331.

Stent, G.S., W.B. Kristan, Jr., W.O. Friesen, C.A. Ort, M. Poon, and R.L. Calabrese. 1978. Neuronal generation of the leech swimming movement. *Science* **200**:1348-1357.

Wrona, F.J., R.W. Davies, L. Linton, and J. Wilkialis. 1981. Competition and coexistence between *Glossiphonia complanata* and *Helobdella stagnalis* (Glossiphoniidae: Hirudinoidea). *Oecologia* **48**:133-137.

Young, J.O. and J.W. Ironmonger. 1979. The natural diet of *Erpobdella octoculata* (L.) (Hirudinea: Erpobdellidae) in British lakes. *Archiv fur Hydrobiologie* **87**:483-503.

3

History of Medicinal Leeching and Early Medical References

Brian Payton
Department of Physiology
Memorial University of Newfoundland
St. Johns, Newfoundland, Canada A1B 3V6

EARLY MEDICAL REFERENCES TO LEECHING

From earliest recorded times, leeches have been used to remove blood from patients. The medicinal use of leeches in ancient India is well-documented. In contrast, references to leeches in classical Greek writing are scarce. Hippocrates (460–377 BC) does not appear to have mentioned the therapeutic use of leeches. A Greek, Nicander of Colophon (c 130 BC), is generally credited with the first Western mention of the medicinal use of leeches. The first printed edition of his work *Theriaca* appeared in 1557 (see Huber 1891). Galen (c 130–201 AD) has been cited as recommending the use of leeches, but the authenticity of this has been questioned. It must be remembered that early works were handed down in manuscript form and often had material added when they were copied. In medieval times, the classical works available were also often Latin translations of Arabic texts, which in turn were translations of the original Greek. Although the use of leeches as a remedy for a variety of maladies was subsequently mentioned by many classical authors, as well as by medieval and Renaissance European authors (see Johnson 1816), it is probably untrue that leeches were an important and frequently used form of therapy in the West until the late 18th century. Even the important and influential medical

FIGURE 1

▲ Illustration from a 16th-century French manuscript by Pierre Boaistuau (1560) showing a man being leeched for obesity. (Courtesy of the Wellcome Trustees.)

treatises by the Persian physician Avicenna (980–1037) (see Gruner 1930) and the Arabian surgeon Albucasis (c 1013), although containing sections giving the mode of application and indications for leeching, devote little space to that procedure.

One of the reasons for the widely held belief that leeches were used intensively in medieval times is that the term "leech" was used then to designate English physicians. Moreover, Anglo-Saxon medical writings were referred to as "leech books" (Cockayne 1864; Payne 1904; Dawson 1934; Wright 1955). These leech books, which were collections of "leechdoms," that is, recipes for treatments, relied heavily on the earlier Greco-Roman sources and on indigenous folk medicine; but they contain hardly any references to leeches or bloodsucking worms. The treatments called leechdoms are predominantly herbal ("wortcunning") and frequently relate to astrological practices ("starcraft").

Of early printed medical treatises, the *Works of Ambroise Paré*, published in French in 1570, must have been one of the most influential. (These works raised the ire of the College of Physicians of Paris, since not only was it against the law to publish medical treatises without the college's approval, as Paré had done, but also such works had to be written in Latin.) The first English translation of Paré's text was printed around 1630 and for well over another 50 years was considered to be definitive. It was a comprehensive work covering anatomy, physiology, alchemy, materia medica, pathology, surgery, medicine, embryology, obstetrics and gynecology, and even much biology, and it was probably the first instance of a multidisciplinary medical text. However, except for a description of their use prior to amputation, leeches are relegated to two short paragraphs in this comprehensive work. One brief paragraph (Book 17, Chapter 52) deals with their identification and mode of application, and the other (Book 21, Chapter 29), placed in a section on poisons, deals with leech infestation. An illustration taken from a 16th century French manuscript and depicting leeches being used for the treatment of obesity is shown in Figure 1.

As for early printed surgical works rather than medical treatises in English, those by William Clowes (1596) and Peter Lowe (1634) mention leeches, but they, too, lack extensive accounts of the use of leeches in bloodletting.

Sixteenth and seventeenth century medicine had a definite hierarchical structure, with practitioners ranking from physicians at the top down to surgeons, barber-surgeons, and apothecaries. Thus, bloodletting was usually delegated to apothecaries and surgeons because physicians found it demeaning to have too close contact with their patients, and those in holy orders were even expressly forbidden to use the knife. One of the few medical treatises in the 17th century devoted to leeching is that by Johannes Heunius (1652).

THEORIES AND METHODS OF LEECHING

The rationale for leeching was based both on the humoral theory of disease and on the concept of counterirritation (Brockbank 1954; Hartnett 1972). The humoral doctrine was an early pathophysiological concept according to which the correct balance of the four humors — blood, phlegm, and yellow and black bile — is altered in disease.

Many therapeutic measures were directed towards restoring the normal balance of these humors. This restoration might be achieved by emetics, purgatives, clysters (enemas), diaphoretics (induce sweating), or bloodletting. Bloodletting could be achieved by venesection (the incision of superficial veins), wet cupping (a technique that involved initial scarification of the skin and subsequent application of cups which were exhausted by burning the air in them so as to create a partial vacuum when applied to the skin), and leeching. Venesection and cupping were much more widely practiced than leeching, but leeching allowed more discrete bloodletting from a wider variety of anatomical sites than the other techniques and was usually considered more suitable for children. Bleeding, whether by venesection, cupping, or leeching, was thought to change not only the total body content of the humors but also their relative concentrations.

Leeching and cupping were also considered as forms of counterirritation, a mechanism probably allied to that held to underlie acupuncture. Other forms of counterirritation were the use of blisters, setons (horse hairs or gauze introduced under the skin), issues (the implantation under the skin of foreign bodies, usually metal, which also formed a site for local infection), cautery, and moxibustion (a treatment whereby small cones of dried mugwort were placed on the skin and then set alight).

LEECHING IN THE 18TH AND 19TH CENTURIES

For reasons now difficult to fathom, a great increase in the practice of bloodletting occurred in the 18th and 19th centuries. Although this increase was more marked in France, the popularity of all methods of bloodletting spread rapidly throughout Europe. The works of Vitet (1809) and Moquin-Tandon (1846) in France, Schmucker (1776) in Germany, and Horn (1798) and Johnson (1816, 1825) in England are examples of the many treatises appearing on leeches around this time.

Francois Victor Joseph Broussais (1722–1838) (Fig. 2) was an influential French physician whose views on pathophysiology identified inflammation of the digestive tract as the basis of almost all diseases (Rolleston 1939). Bloodletting, particularly by leeching, was an important feature of his treatments, and he is said to have ordered 30 leeches for each of the patients in his clinics even before he saw them. Not all physicians showed this sanguinary approach, however. Pierre Charles Alexandre Louis (1787–1872), an eminent physician and colleague of Laënnec, was one of the earliest to apply statistical methods to assess the value of various therapies. As early as 1820, he published

▲ Early 19th-century lithograph by C.T. de Villers (1804–1859) depicting F.J.V. Broussais. The caption reads, "Another ninety leeches and continue with the diet." (Courtesy of the Wellcome Trustees.)

data showing that bloodletting was harmful rather than beneficial (Louis 1836).

Possibly the ease with which bloodletting, by leeching or any other method, could be carried out helped its popularity and led to abuse. The Edinburgh physician William Buchan was the author of a successful home medicine text (Buchan 1788) that ran to many editions and contains much sensible advice. It includes two indications for the use of leeches — one is intractable pain during teething in infants, and the other is best put in his own words: "I have been of use for some time past, to apply leeches to inflamed testicles, which practice has always been followed with the most happy effects."

The physician who contributed the home medicine section to the popular English best-seller *The Book of Household Management* by Mrs. Isabella Beeton (1861) was either unaware of Pierre Louis' findings or opposed to them, since one of his most frequent recommendations was to engage in various acts of exsanguination should a member of the household fall sick.

The height of abuse of leeching appears to be the reported attempt of a young woman to commit suicide by the use of leeches (Suicide 1892). Unfortunately, the published account of this case did not indicate whether the attempt was successful or not; however, according to Broussais' teaching, the 50 leeches the woman applied to herself should have prolonged her life rather than killed her.

The 19th century mania for leeching was not limited to Europe, and Hartnett's (1972) article describes a similar phenomenon on the North American scene. In America, the New World species *Macrobdella decora* was often used for leeching, but the traditional Old World species *Hirudo medicinalis* was still considered to be superior, and large numbers of specimens were consequently imported from Europe.

Medicinal leeching reached its peak between 1825 and 1840; its subsequent decline by the end of the 19th century can be appreciated by comparing two widely used medical textbooks. Thomas' *Modern Practice of Physic* (1825), which was written in the heyday of leeching, has indications for leeches in very many conditions and for venesection in still more. Thomas also says: "To trust wholly to leeches without general bleeding, is only tampering with a most formidable disorder." In contrast, the first edition of the forerunner of modern medical texts, William Osler's, *Principles and Practice of Medicine* (1892), has only 11 references to bloodletting of any sort. Three of these are probably justifiable by present-day standards in the absence of drugs. In two of these three references to leeches, he suggests that an injection of morphine is preferable.

CONTEMPORARY USE OF LEECHES

Apart from a few scattered references, therapies based on the application of leeches have appeared only rarely in contemporary medical literature. However, scientific interest has continued into the 20th century (Mann 1962). The complete bibliography published by Autrum (1939) lists over 2500 research papers on leeches. The discovery of the secretion of the anticoagulant hirudin by the leech (Haycraft 1884) and its subsequent identification and the elucidation of its mechanisms of action by Markwardt (1957) have not only been of medical interest for anticoagulant therapies, but have been important for elucidating the mechanism of blood clotting. The leech body wall also had an important use from 1932 until fairly recently in a bioassay technique for acetylcholine (Minz 1932; Szerb 1960). Current interest in the neurobiology of the leech is now making it of even wider significance.

REFERENCES

Albucassis. c 1013. *On surgery and instruments.* (Translated by M.S. Spink and G.L. Lewis, 1973.) University of California Press, Berkeley and Los Angeles.

Autrum, H. 1939. Literatur uber Hirudineen. In *Klassen* and *Ordnungen des Tierreichs*, (ed H.G. Bronns), 4, III, 4 Hirudineen.

Beeton, I. 1861. *The book of household management.* (Facsimile, 1969.) Cape, London.

Boaistuau, P. 1560. *Histoires prodigieuses.* Wellcome M.S., London, WHMM 6573.

Brockbank, W. 1954. *Ancient therapeutic arts.* Heinemann, London.

Buchan, W. 1788. *Domestic medicine*, 10th Edition. A. Strahan (printers), London.

Clowes, W. 1596. *A profitable and necessary book of observations.* Scholars' Facsimiles and Reprints, New York.

Cockayne, O. 1864. *Leechdoms, wortcunning and starcraft*, vol. 1–3. Revised ed. (Introduction by C. Singer, 1961.) Holland Press, London.

Dawson, W.R. 1934. *A leechbook on collection of medical recipes of the 15th century.* London.

Gruner, O.C. 1930. *A treatise on the Canon of Avicenna, incorporating a translation of the first book.* Luzac, London.

Hartnett, J.C. 1972. The care and use of medicinal leeches in 19th century pharmacy and therapeutics. *Pharm. Hist.* **14**:127–138.

Haycraft, J.B. 1884. On the action of a secretion obtained from the medicinal leech on the coagulation of blood. *Proc. R. Soc.* **36**:478–487.

Heunius, J. 1652. *De hirudinum usu et efficacia in medicina tractatus.* J. Jegeri, Gryphiswaldiae.

Horn, G. 1798. *An entire new treatise on leeches wherein the singular and valuable reptile is most clearly set forth.* London.

Huber, J. 1891. *Nicander — Theriaca.* apud Guil. Morelium. *Dtsch Arch. Klin Med.* **67**:522.

Johnson, J.R. 1816. *A treatise on the medicinal leech.* Longman, Hurst, Rees, Orme and Browne, London.

———. 1825. *Further observations on the medicinal leech.* Longman, Hurst, Rees, Orme, Browne and Green. London.

Louis, P.C.A. 1836. *Researches on the effect of blood letting in some inflammatory diseases and on the influence of tartarised antimony and vesication in pneumonias* (Translated by C.G. Putnam). Hilliard Gray, Boston.

Lowe, P. 1634. *A discourse on the whole art of chirurgerie.* Book 9, chapter 13, pp.

391–393. T. Purfoot, London.
Mann, K.H. 1962. *Leeches (Hirudinea). Their structure, physiology, ecology and embryology.* Pergamon Press, New York.
Markwardt, F. 1957. Die Isolierung und Chemische Charakterisirung des Hirudin. *Hoppe-Seyler's Z. Physiol. Chem.* **308**:147–156.
Minz, B. 1932. Pharmakologische Untersuchungen am Blutegelpraeparat: Zugleich eine Methode zum biologischen Nachweis von Acetylcholine bei Arwesenheit anderer pharmakologisch wirksamer koerpereigner Stoffe. *Naunyn-Schmeiderbergs Arch. Pathol. Pharmakol.* **168**:292–304.
Moquin-Tandon, A. 1846. *Monographie sur la famille des hirudinees, atlas.* J.B. Bailliere, Paris.
Osler, W. 1892. *The principle and practice of medicine.* Young and Pertland, London.
Paré, A. 1678. *The works of Ambroise Paré*, 2nd edition. (Translated by Johnson.) M. Clark (printer), London.
Payne, J.F. 1904. *English medicine in Anglo-Saxon times.* Clarendon, Oxford.
Rolleston, J.D. 1939. F.J.V. Broussais (1771-1838): His life and doctrines. *Proc. R. Soc. Med.* **32**:405–413.
Schmucker, J.L. 1776. Historisch-practische Abhandlung vom medicinischen Gebrauche der Blutegel. *Verm. Chir. Schrift. Berl.* **1**:75–116.
Suicide. 1892. Note in *N.Y. Med. J.* **56**:103.
Szerb, J.C. 1961. The estimation of acetylcholine using leech muscle in a microbath. *J. Physiol.* **158**:8P–9P.
Thomas, R. 1825. *The modern practice of physic*, 8th edition. Collins and Collins, New York.
Vitet, L. 1809. *Traite de la sangsue medicinale.* Vitet, Paris.

4
Structure of the Leech Nervous System

Brian Payton
Department of Physiology
Memorial University of Newfoundland
St. Johns, Newfoundland, Canada A1B 3V6

The following description of the gross anatomy of the leech is primarily intended to provide a background against which the location and organization of the nervous system can be appreciated. (For more detailed descriptions and reference sources see Mann [1962] and Harant and Grassé [1959].) Since many of the neurobiological studies have been carried out on *Hirudo medicinalis*, only this species is described. In many respects, the organizations of *Macrobdella decora* and *Haemopis marmorata* (family Hirudinae) are similar to that of *H. medicinalis*.

MORPHOLOGICAL FEATURES OF *H. MEDICINALIS*

As might be expected, the size, shape, and weight of an adult specimen of *H. medicinalis* depend on its state of feeding and activity. Adult leeches can weigh 2–3 g, but because they ingest 5–10 ml of blood in one feeding, their weights are highly variable. When at rest and fixed only by its tail sucker, the animal varies in length from 5 to 10 cm, and, in this position, the head and rostral half of the body are narrower than the caudal half. When swimming, the leech's body flattens dorsoventrally, and the body can stretch to 15 or 20 cm. On

FIGURE 1

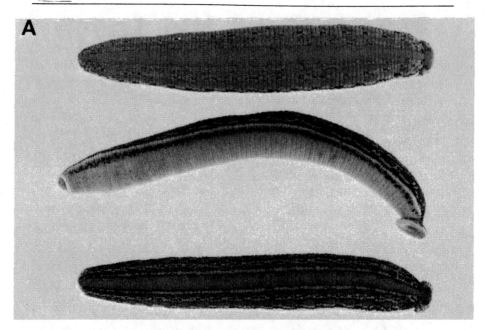
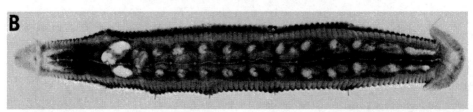
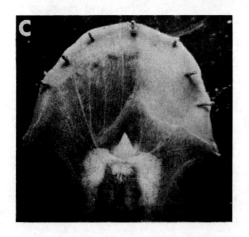
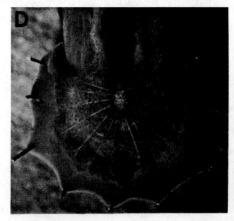

being disturbed, its body adopts an almost globular shape. Anaesthetized leeches are shown in Figure 1A. The upper specimen in this figure was 7 cm long and 1.3 cm wide.

The details of the metameric segmentation of leeches have been the subject of some dispute (Mann 1953). Nevertheless, it is evident that of the 102 annuli that are clearly visible on its surface, the first 10 annuli relate to the first 4 segments of the head region (Yau 1976). The next, or first body segment, has 3 annuli, the second body segment has 4 annuli, and the next 15 body segments have 5 annuli each. Small, discrete spots that are sense organs, or sensilla, are found on one annulus in each segment. In the midbody segments, there are seven pairs of these sensilla distributed over both the dorsal and ventral surfaces. In such body segments, this annulus marks the center of the segment and the level of the corresponding nerve ganglion, which will be described later. These sensilla have different locations in the head and in the first and last 4 body segments (Mann 1953; Yau 1976), because the numbers of annuli per segment decrease here. The sensilla contain mechanoreceptors (Derosa and Friesen 1981) and photoreceptors; in the head region, there are five pigmented pairs of "eyes" on the dorsal surface (Kretz et al. 1976).

The numbering of the leech body segments is subject to some confusion because anterior to the frontmost metameric segment lies a prostomium that should not be included in the counting process

◄ *H. medicinalis.* (*A*) Anaesthetized leeches. Upper leech is 7 cm long. Note the different patterns of dorsal pigmentation in the upper and lower specimens. The male pore is clearly visible on the ventral surface of the middle specimen. The female pore is just visible 5 annuli caudally. (*B*) The ventral midline has been incised and reflected laterally. The epididymi and penis sheath lie adjacent to and just below the fifth nerve ganglion. The ovisacs and vagina lie below the sixth ganglion. Nine pairs of testicles are located within segments 7 through 15. The ventral blood sinus has been opened to expose the chain of nerve ganglia and their connectives. (*C*) A ventral dissection of the head region. One dorsal and two ventrolateral jaws surround the entrance to the pharynx. Arising from the subesophageal ganglion are three pairs of ventral nerves. On the right side of the head, two of the four paired dorsal nerves have been exposed by removing the right velum and skin of the mouth cavity. The first and second body ganglia are still enclosed within the ventral blood sinus. (*D*) Ventral dissection of the tail. Seven pairs of nerves can be seen radiating out from the tail ganglion. The last three body ganglia are visible within the ventral blood sinus.

(Mann 1962). For neurobiological purposes, it is more convenient to describe the segmentation according to head and body segments (Kristan et al. 1974a), the frontmost 4 segments being designated as head segments 1–4 and the following 21 segments as body segments 1–21, with the 7 caudal segments being fused to form the tail sucker. This system for numbering body segments is used throughout this book and is illustrated in Figure 2.

The pattern of skin pigmentation in *H. medicinalis* is markedly differ-

FIGURE 2

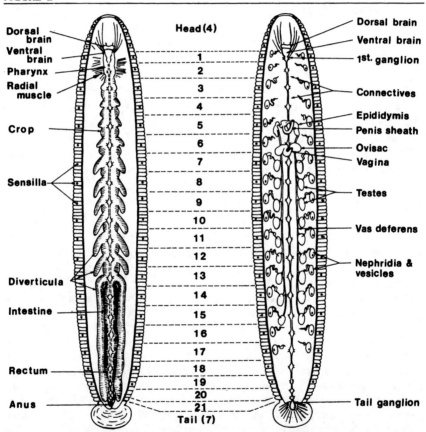

▲ Segmentation and the positions of digestive tract, sex organs, nephridia, and nerve cord in *H. medicinalis*. The nerve cord lies ventrally and the anus opens onto the ventral surface. The supraesophageal ganglion is designated as "dorsal brain" and the subesophageal ganglion as "ventral brain" in this diagram. (Redrawn from Mann 1962.)

ent on the dorsal and ventral surfaces. The ventral surface is usually light green, and a lateral stripe of black pigmentation demarcates the ventral and dorsal surfaces. The darker green dorsal surface may show some variation, as illustrated in Figure 1A. Generally, there are two broad, orange stripes coursing rostrocaudally on the dorsolateral surface, with an additional thinner orange stripe sometimes lying more laterally. Lines of black pigment are associated with these orange-pigmented stripes.

In forming the anterior sucker in the head region, the annuli on the ventral surface are distorted, so that the dorsal surface projects more anteriorly. Inside the anterior sucker lies the sucker cavity; this is separated from the buccal cavity of the alimentary canal by a fold of skin, the velum. Underlying the velum and projecting into the buccal cavity lie three jaws that radiate out from the opening leading into the pharynx (Fig. 1C). A ridge of modified chitin on the surface of these jaws forms a row of small teeth. The rocking action of these jaws cuts through the host's skin after the leech has attached itself by means of its anterior sucker. The marks left by a leech bite appear as three radiating lines; in fact, William Harvey (1628) likened the apposition of the cusps of the tricuspid valve to a leech bite.

The pharynx courses through body segments 1 and 2 and connects with the crop. The distinction between pharynx and esophagus is somewhat arbitrary (see Fig. 2). As the crop extends through body segments 3–12, it gives rise to lateral diverticula. These become progressively larger as the crop extends caudally. In a transverse section of the rear of the body (Fig. 3), the diverticulum arising in body segment 13 can be seen to course parallel to the continuation of the alimentary canal. The rectum exits at the anus, which is situated on the ventral surface just rostral to the tail sucker. Numerous muscle fibers, which presumably are involved in the sucking action during the ingestion of blood, radiate out from the pharynx to the body wall. Apart from secretory glands in segments 1 and 2, which open into the pharynx, the digestive system is not well differentiated. Digestive processes are carried out by bacteria (*Pseudomonas hirudinae*) that live in symbiosis with the leech (Büsing 1951; Büsing et al. 1953). The salivary secretion contains an anticoagulant, hirudin (Haycraft 1884; Yanagisawa and Yokoi 1938), which prevents clotting of the blood ingested by the leech. In addition, a histaminelike compound is injected into the host's tissue; vasodilation caused by this material is thought to account for the prolonged bleeding following a leech bite. The amount of blood loss from a leech bite is frequently as much as 20–50 ml, as the site may continue to bleed for 24 hours (Derganc and Zdravic 1960).

FIGURE 3

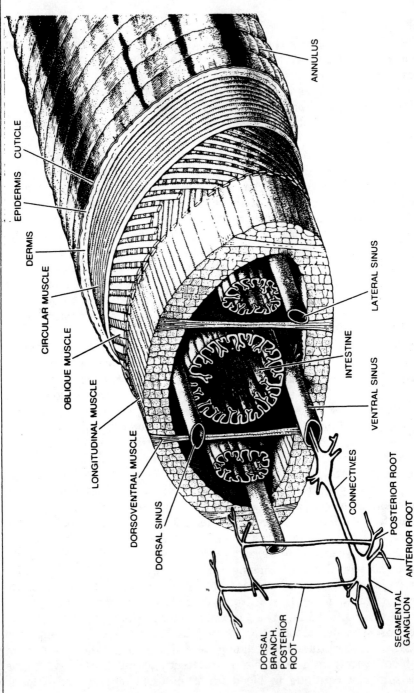

▲ Diagram to show the position of the nerve cord within the ventral blood sinus and its relation to other body structures and the musculature. (Reprinted, with permission, from Nicholls and Van Essen 1974.)

Body segments 5 and 6 contain the sexual organs. The epididymi (the large white areas in Fig. 1B), the prostate, and the penis sheath are located in segment 5. The ovaries, oviductal gland, and vagina are immediately caudal to these and cross anteriorally to the nerve cord. The male pore opens between the third and fourth annulus of segment 5 and is usually obvious (Fig. 1A). The female pore is less distinct and lies between the fourth and fifth annulus of segment 6. The penis is eversible from the penile sheath (Zipser 1979) and during copulation is inserted into the vagina of another leech. As Zipser (1979) rightly points out in her study on the control of penile eversion, "it [copulation] must tax the intellectual capabilities of the 15,000- to 20,000-neuron central nervous system." There are nine or ten pairs of testes occupying body segments 7 to 15 or 16 (there are 9 pairs in Fig. 1B). The testes communicate with the epididymi via a pair of fine vasa deferentia which extend rostrally.

The excretory system is made up of 17 pairs of nephridia located in body segments 1–17. The nephridia communicate with the outside through a pair of nephridiopores, which open onto the ventral surface in the groove that lies immediately in front of the annulus containing the sensilla of that segment.

The vascular system has four major longitudinal sinuses—one dorsal sinus, one ventral sinus, and a pair of lateral sinuses. The lateral sinuses function as the heart tubes, and their rhythmic constriction pumps blood through the vascular system (Thompson and Stent 1976). Although the detailed circulation pattern has not been fully described, blood from the lateral sinuses appears to circulate via its branches to the body wall and through the botryoidal tissue, returning via the ventral and dorsal sinuses (Gratiolet 1862; Lankester 1880; Bradbury 1959). (The botryoidal tissue is a conspicuous loose network of tissue containing pigmented cells and vascular channels and is visible in Figure 1D.)

The body wall is mainly made up of three layers of muscle. From the outside inwards, these layers have muscle fibers that course circumferentially, obliquely, and longitudinally, the innermost longitudinal layer being the thickest (Stuart 1970). Just beneath the skin are additional muscle fibers that extend rostrocaudally for the length of an annulus. When they contract, they cause each annulus to form a sharp ridge and are therefore known as annulus erector muscles. Dorsoventral muscle fibers traverse the body cavity, as depicted in Figure 3; their contraction flattens the body. Circular, radial, and longitudinal muscle fibers can also be found in the mouth and tail suckers, but the layers are not so clearly discernable.

GENERAL ORGANIZATION OF THE NERVOUS SYSTEM

The 32 metameric segments of the leech plus the nonsegmental prostomium are innervated by a head ganglion, 21 body ganglia, and 7 fused tail ganglia (Fig. 2). These ganglia form a chain that lies on the ventral aspect of the alimentary canal, from annuli 11, 12, and 13 to the tail sucker. The ganglia are joined by connectives that consist of two large lateral bundles of nerve fibers and a thin medial connective called Faivre's nerve (Fig. 4). The connectives between the head ganglion and the first body ganglion, as well as those between the last body ganglion and the first tail ganglion, are extremely short. The nerve cord is covered by a fibrous sheath or capsule; in the connectives, this sheath contains contractile muscle fibers.

The whole of the nerve cord is enclosed within the ventral blood sinus (Fig. 3), which is closely attached to the head and tail ganglia. The ganglia and connectives lie free in the body segments, but there are no blood vessels within the central nervous system (CNS). Owing to this feature and the small size of the nerve cord, materials move rapidly through a ganglion. For example, the half time for exchange of molecules such as sucrose or small ions is of the order of seconds (Kuffler and Nicholls 1966).

A pair of nerve roots arise from each side of the body ganglion and branch to innervate segmental structures in that region. The posterior root gives off a dorsal branch that courses around the alimentary tract and innervates the dorsal surface. A nomenclature and system for describing these branches is given by Ort et al. (1974). The peripheral nerves form complex networks consisting of nerve rings and longitudinal tracts (Livanov 1904). There are also peripheral sensory cells (see Chapter 5), some of which have large axons that enter the ganglion. However, there are no peripheral reflexes that contribute to body movements, and all responses to mechanical stimuli disappear after removal of ganglia. In this respect, leeches differ from molluscs.

THE CELLULAR ORGANIZATION OF GANGLIA IN THE VENTRAL NERVE CORD

Each segmental ganglion contains about 400 nerve cell bodies, with some slight variation from ganglion to ganglion (Macagno 1980). Ganglia 5 and 6, which innervate the sex organs, contain many more cells (over 700). The neurons are monopolar and their cell bodies, whose diameters range from $10\mu m$ to $60\mu m$, are separated by septa into six groups, known as packets. Their axonal processes and neuritic

FIGURE 4

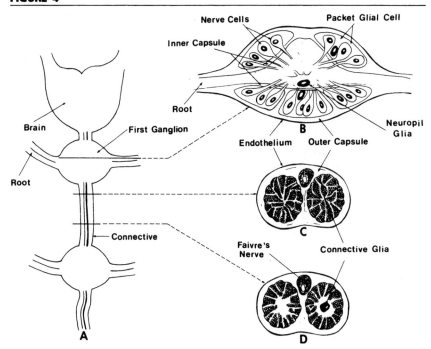

▲ Diagrams of the leech CNS. (A) The head ganglion (brain) and first two body ganglia. (B,C,D) Sections of the nerve cord (dorsal side up) to show the relation of the large glial cells to other structures. The endothelium is a complete cellular lining covering the outer capsule. The glial nucleus is indicated in only two of the packets. (Redrawn from Kuffler and Potter 1964.)

trees are contained within, or pass through, a central neuropil (see Fig. 4B), (see Coggeshall and Fawcett [1964] for detailed ultrastructure and early morphological references). A single glial cell envelops the neurons in each packet, and two more glial cells surround the processes in the neuropil. In addition, two more glial cells are present, one in each lateral connective.

The general appearance of ganglia viewed from the ventral aspect and by transmitted light is shown in Figure 5B. The appearance and relationships of the cells in individual ganglia are not only similar from ganglion to ganglion but also from one leech to another. The ganglia of leeches belonging to the same suborder may also appear very similar (cf. Fig. 5 B and C). Two large neuronal cell bodies, the Retzius (1891), or *kolassel*, cells are usually clearly visible in the anteromedial packet. A

FIGURE 5

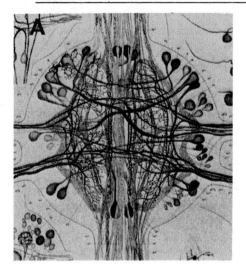
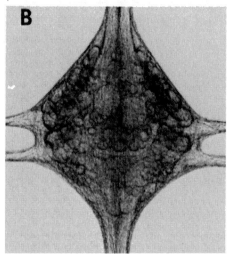
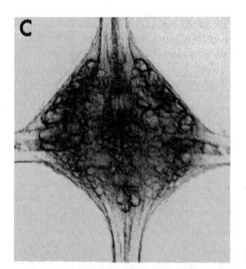
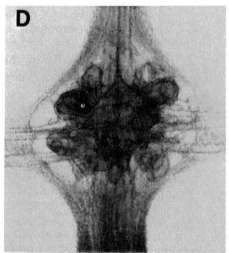

▲ Leech ganglia viewed from the ventral surface. (*A*) Retzius' (1891) drawing, *H. medicinalis*. (*B*) Photograph, *H. medicinalis*. (*C*) *Haemopis marmorata*. (*D*) *Haementeria ghilianii*. Note the similar appearance of *B* and *C*. These are both gnathobdellid leeches; *D* is a giant rhynchobdellid leech and has fewer cells than *B* and *C*.

diagram illustrating the organization of the cell packets within a ganglion, viewed from the dorsal and ventral surfaces, is shown in Figure 6. It should be noted that the septa, which dip down into the ganglia to delineate the packets, slope in various directions; therefore, the outline of a packet depends on the actual level within the ganglion. Macagno (1980) has devised a numbering system for these packets and has also studied the variations in the numbers of cells within the equivalent packets of different ganglia; the numbers are very similar, but small variations do occur.

Physiological Properties of Glial Cells

The large size of the glial cells and their special relationship to the neurons makes them amenable to intracellular recording and other physiological techniques (Kuffler and Potter 1964; Kuffler and Nicholls 1966). The resting potential of glial cells, like that of neurons, is largely governed by the K^+ concentrations, but glial cells have higher resting potentials (-75 mV compared with -50 mV). Neurons can also signal electrically when the glia are removed. Glial cells do not appear to be directly involved in immediate signaling processes, but, as in the mammalian brain, impulses in neurons lead to depolarization of glia by means of K^+ accumulation in the narrow extracellular spaces (see also Nicholls and Baylor 1968a; Baylor and Nicholls 1969). The glial cells are electrically coupled to each other, and, at least over short periods of time, do not appear to be necessary for the transport of electrolytes or glucose from the extracellular space to the neurons (Nicholls and Wolfe 1967; Wolfe and Nicholls 1967).

Organization of Neurons

The remarkably stereotyped morphological appearance, together with the large size of the neuron cell bodies, makes the segmental ganglia ideal preparations for identifying the functions of individual cells and for studying their physiological properties. The first neurons to be identified functionally were the sensory neurons serving touch, pressure, and nociception (Nicholls and Baylor 1968b) — the T, P, and N cells. In addition, the motor cells and a variety of interneurons important for swimming, heartbeat, and sexual responses of the animal are now known and are described in detail in succeeding chapters. Often, leech nerve cells can be identified tentatively by simple visual inspection; then the diagnosis can be confirmed by recording intracellularly. A number of the cells of unknown function, such as the Leydig cell, are also identifiable (see Appendix C).

FIGURE 6

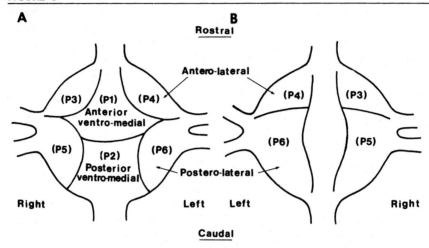

▲ Diagram of the six cell packets of a ganglion as seen when viewed from either the ventral (A) or dorsal (B) surface. The actual outline of each packet will depend on the exact plane within the ganglion. The neuropil lies dorsomedially and has two glial cells. P1 through P6 is the packet notation given by Macagno (1980).

THE HEAD GANGLION

The neurons of the head ganglion have not received the same attention as the neurons of the body ganglia, perhaps because they are less easily accessible and more complex than the body ganglia. The head ganglion is located beneath annuli 11, 12, and 13, and consists of two main masses. The mass above the pharynx, the supraesophageal ganglion, is formed from a cluster of cells lying in front of the larval aperture (J. Fernandez, pers. comm.). The mass below the pharynx, the subesophageal ganglion, is formed by fusion of the four neuromeres of the rostral four metameric segments. (The German terminology for these ganglia is *Ober-* and *Unter-Schlundganglienmasse*, thus indicating the position relative to the pharynx [*Schlund*]. Writers in the English language have used the terms supra- and sub- "pharyngeal" or "esophageal" interchangeably; some authors have used both terms in the same paper. The more common usage now appears to be "esophageal.") Two periesophageal commisures connect the supra- and subesophageal ganglia (Fig. 7).

The subesophageal ganglion, which is shown in Figure 7, was studied by Yau (1976) to ascertain whether the organization of mecha-

FIGURE 7

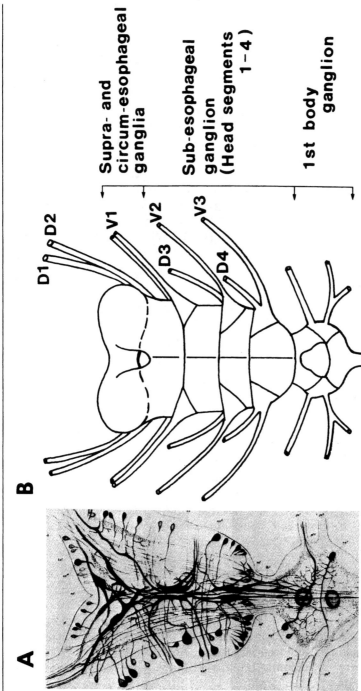

▲ The leech head ganglion. (A) Retzius' (1891) drawing of the subesophageal, or subpharyngeal, ganglion. (B) Diagram to show the segmental organization and the origins of the dorsal and ventral nerves. Nerves D1–D4 course dorsal to the ventrolateral jaws; nerves V1–V3 course ventral to them.

noreceptor cells in its four neuromeres is similar to that found in the body ganglia. He was able to identify and map the receptive fields of mechanoreceptor neurons similar to those found in body ganglia by Nicholls and Baylor (1968b). The nerves arising from the subesophageal ganglion have a complex branching pattern and distribution (Yau 1976), which have also been charted by Kretz et al. (1976) in a study of photosensory input pathways. The first pair of dorsal nerves arise from the junction of the periesophageal commisure and the supraesophageal ganglion (Fig. 1C and Fig. 7), and they extend to innervate the more dorsal and anterior parts of the anterior sucker and the first few annuli. Three other paired nerves extend dorsally; the second pair arises from the periesophageal commisure, and the other two pairs arise from more dorsal regions of the subesophageal ganglion. These four pairs of dorsal nerves all course dorsally to the two ventrolateral jaws. Three pairs of ventral nerves arise from the subesophageal ganglion. The first pair arises more anteriorly and soon forms two major branches, which course to the anterolateral parts of the sucker. The next two pairs extend to the skin around the lateral and ventrolateral parts of the sucker, respectively. In Figure 1C, the ventral midline of the sucker has been reflected and pinned laterally. The lateral displacement of these nerves has also influenced the diagram in Figure 7.

The mechanoreceptor cells in the subesophageal ganglion studied by Yau do not send extensive branches into the dorsal brain. The supraesophageal ganglion of *Macrobdella decora* has been investigated by Orchard and Webb (1980). They found neurons with axons and dendritic trees, but the processes of these neurons appeared to be limited to within the supraesophageal ganglion. The nature of the processes and their association with vascular channels suggested a prominent neurosecretory role for cells in this region (see also Appendix B).

A circumpharyngeal "sympathetic" nerve ring has also been described as lying just anterior to the subesophageal ganglion (Livanov 1904), but nothing is known of its function.

THE TAIL GANGLION

The tail ganglion is formed by fusion of the seven neuromeres of the caudal seven metameric segments. Its neurons have not been systematically investigated. Seven pairs of nerves arise from the tail ganglion and radiate out to innervate the tail sucker (Rubin 1978). In some preparations, a fine additional nerve lying in the midline may be visible.

REFERENCES

Baylor, D.A. and J.G. Nicholls. 1969. Changes in extracellular potassium concentration produced by neuronal activity in the central nervous system of the leech. *J. Physiol.* **203**:555–569.

Bradbury, S. 1959. The botryoidal and vasofibrous tissue of the leech, *Hirudo medicinalis*. *Q. J. Microsc. Sci.* **100**:483–498.

Büsing, K.H. 1951. *Pseudomonas hirudinis* ein baktrieller Darmsymbiont des Blutegels (*Hirudo medicinalis*). *Zentralbl. Bakteriol. I Orig.* **157**:478–484.

Büsing, K.H., W. Döll, and K. Freytag. 1953. Die Bakterienflora del medizinischen Blutegel. *Arch. Mikrobiol.* **19**:52–86.

Coggeshall, R.E. and D.W. Fawcett. 1964. The fine structure of the central nervous system of the leech, *Hirudo medicinalis*. *J. Neurophysiol.* **27**:229–289.

Derganc, M. and F. Zdravic. 1960. Venous congestion of flaps treated by application of leeches. *Brit. J. Plast. Surg.* **13**:187–192.

Derosa, Y.S. and W.O. Friesen. 1981. Morphology of leech sensilla: Observations with the scanning electron microscope. *Biol. Bull.* **160**:(in press).

Gratiolet, P. 1862. Recherches sur l'organisation du système vasculaire dans la sangsue medicinale et l'aulastome vorace. *Ann. Sci. Nat. Zool.* **17**:174–225.

Harant, H. and P.-P. Grassé. 1959. Classe des annélidees achètes ou hirudinées ou sangsues. In *Traité de zoologie*. (ed. P.-P. Grassé), vol. 5, pp. 471–595. Masson, Paris.

Harvey, W. 1628. *De Moku Cordis*. G. Fitzeri, Frankfurt.

Haycraft, J.B. 1884. On the action of a secretion obtained from the medicinal leech on the coagulation of blood. *Proc. Roy. Soc.* **36**:478–487.

Kristan, W.B., G.S. Stent, and C.A. Ort. 1974. Neuronal control of swimming in the medicinal leech. I. Dynamics of the swimming rhythm. *J. Comp. Physiol.* **94**:97–119.

Kretz, J.R., G.S. Stent, and W.B. Kristan. 1976. Photosensory input pathways in the medicinal leech. *J. Comp. Physiol.* **106**:1–37.

Kuffler, S.W. and J.G. Nicholls. 1966. The physiology of neuroglial cells. *Ergeb. Physiol. Biol. Chem. Exp. Pharmakol.* **57**:1–90.

Kuffler, S.W. and D.D. Potter. 1964. Glia in the leech nervous system: Physiological properties and neuron-glia relationships. *J. Neurophysiol.* **27**:290–320.

Lankester, E.R. 1880. On the connective and vasifactive tissue of the medicinal leech. *Q. J. Microsc. Sci.* **20**:307–317.

Livanov, N. 1904. Untersuchungen zur Morphologie der Hirudineen. II. Das Nervensystem des Vorderen Körperendes und seine Metamerie. *Zool. Jahrb. Anat.* **20**:153–226.

Macagno, E.R. 1980. Number and distribution of neurons in the leech segmental ganglion. *J. Comp. Neurol.* **190**:283–302.

Mann, K.H. 1953. The segmentation of leeches. *Biol. Rev.* **28**:1–15.

———. 1962. *Leeches (Hirudinea). Their structure, physiology, ecology and embryology.* Pergamon, Macmillan, New York.

Nicholls, J.G. and D.A. Baylor. 1968a. Long lasting hyperpolarization after activity of neurons in the leech central nervous system. *Science* **162**:279–281.

———. 1968b. Specific modalities and receptive fields of sensory neurons in the CNS of the leech. *J. Neurophysiol.* **31**:740–756.

Nicholls, J.G. and D. Van Essen. 1974. The nervous system of the leech. *Sci. Amer.* **230**:38–48.

Nicholls, J.G. and D.E. Wolfe. 1967. Distribution of ^{14}C-labeled sucrose, insulin, and dextran in extracellular spaces and in cells of the leech central nervous system. *J. Neurophysiol.* **30**:1574–1592.

Orchard, I. and R. Webb. 1980. The projections of neurosecretory cells in the brain of the North-American medicinal leech, *Macrobdella decora*, using intracellular injection of horseradish peroxidase. *J. Neurobiol.* **11**:229–242.

Ort, C.A., W.B. Kristan, and G.S. Stent. 1974. Neuronal control of swimming in the medicinal leech. II. Identification and connections of motor neurons. *J. Comp. Physiol.* **94**:121–154.

Retzius, G. 1891. Zur Kenntniss des centralen Nervensystems der Wurmer. *Biologische Untersuchungen, Neue Folge II*, 1–28. Samson & Wallin, Stockholm.

Rubin, E. 1978. The caudal ganglion of the leech, with particular reference to homologues of segmental touch receptors. *J. Neurophysiol.* **9**:393–405.

Stuart, A.E. 1970. Physiological and morphological properties of motoneurones in the central nervous system of the leech. *J. Physiol.* **209**:627–646.

Thompson, W. and G.S. Stent. 1976. Neuronal control of heartbeat in the medicinal leech. I. Generation of the vascular constriction rhythm by heart motor neurons. *J. Comp. Physiol.* **111**:261–279.

Wolfe, D.E. and J.G. Nicholls. 1967. Uptake of radioactive glucose and its conversion to glycogen by neurons and glial cells in the leech central nervous system. *J. Neurophysiol.* **30**:1593–1609.

Yau, K.-W. 1976. Physiological properties and receptive fields of mechanosensory neurones in the head ganglion of the leech: Comparison with homologous cells in segmental ganglion. *J. Physiol.* **263**:489–512.

Yanagisawa, H. and E. Yokoi. 1938. The purification of hirudin and action principle of *Hirudo medicinalis*. *Proc. Imp. Acad. Tokyo.* **14**:69–70.

Zipser, B. 1979. Identifiable neurons controlling penile eversion in the leech. *J. Neurophysiol.* **42**:455–464.

5
Sensory Cells and Motor Neurons

Susanna E. Blackshaw
Institute of Physiology
University of Glasgow
Glasgow, Scotland G12 8QQ

In the leech, a major advantage for studying behavioral responses and analyzing synaptic interactions within the central nervous system (CNS) has been the rigorous and extensive definition of the sensory input and the motor outflow. Each individual cell that plays a part in the performance of a reflex movement can often be characterized in detail; this is not the case in numerous other invertebrates and vertebrates, where many neurons are involved. Thus, on one side of a ganglion, a single motor cell is responsible for raising all the annuli, a single motor cell innervates a well-defined group of circular muscle fibers, and only two cells respond to pressure applied to the skin. This simplicity and economy make it possible to describe in detail for mechanosensory cells: (1) the specific modality, the response characteristics, and the electrical properties; (2) the branching pattern of the axons in the CNS and in the periphery; (3) the receptive field organization; (4) the morphology of sensory endings; (5) the effects of denervation and reinnervation; and (6) the central connections. A comparable analysis has also been made for motor cells.

It has become apparent that in the head ganglion of *Hirudo* (Yau 1976a) and in the segmental ganglia of widely different species of leeches, such as *Haementeria* (Kramer and Goldman 1981), similar cells with similar properties play corresponding roles. This reinforces the impression that solving the circuitry of a segmental ganglion could provide clues for understanding complex responses of the animal as a whole.

IDENTIFICATION OF SENSORY AND MOTOR NERVE CELLS IN LEECH GANGLIA

By simple visual inspection, as Retzius (1891) and others had shown earlier, it is possible to identify individual cells reliably in leech ganglia according to their shapes, sizes, and positions. The identification can be confirmed by recording electrically from individual neurons with fine intracellular microelectrodes (Nicholls and Baylor 1968). This provides unambiguous criteria for recognizing them in individual preparations. For example, the three groups of cells — touch (T), pressure (P), and nociceptive (N) (Fig. 1) — are all sensory and respond to touch, to pressure, or to noxious mechanical stimulation of the skin. The annulus erector (AE) cells are motor neurons that cause contraction of specific groups of muscles in the body wall. Their structure and physiological functions will be described more fully below. Examples of intracellular records are shown in Figure 1. The impulses in the cell body of T cells are always similar but are smaller and briefer than those in P or N cells; N cells have a particularly large undershoot. In motor nerve cells, the impulses and electrical properties are again different and distinctive. With practice, it is usually enough just to look at a single action potential to be certain which particular cell the electrode is in. Recently, Salzberg et al. (1973; see also Grinvald et al. 1981) have devised an optical technique for recording an impulse from a leech cell without using electrodes. Instead, one measures fluorescence changes in a single cell in a ganglion bathed in saline containing the dye merocyanin. An example is shown in Figure 2. Eventually, it may become possible to scan the activity in groups of cells in this way.

How can one establish the function of a particular cell in the ganglion? A convenient preparation is shown in Figure 3. A part of the body wall of the leech is removed together with the ganglion innervating it. The initial procedure for determining the role of a cell is to record from it while various stimuli are applied to the skin. If it gives impulses, it may well be a sensory cell. If, on the other hand, one stimulates the cell and sees that contraction of a group of muscle fibers follows, one suspects that the cell may be a motor neuron. These procedures are useful starting points, although they do not on their own provide critical evidence of the function of the neuron.

Figure 4 illustrates the response of the sensory cells to various forms of cutaneous stimuli. The T cells fire in response to light touch of the

FIGURE 1

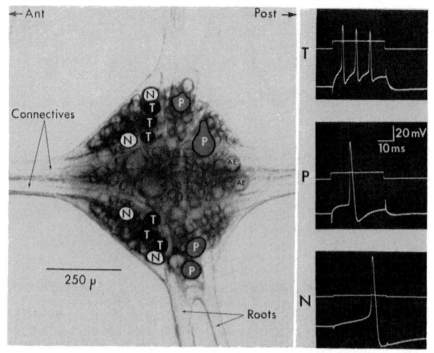

▲ (*Left*) Segmental ganglion seen from its ventral aspect and action potentials of identified cells. Cells labeled include the T, P, and N mechanosensory cells, which respond to touch, pressure, or noxious mechanical stimulation of the skin, and the AE motor neurons, which innervate the muscles responsible for raising the annuli into ridges. (*Right*) The intracellular recordings of T, P, and N action potentials were elicited by passing depolarizing current through the microelectrode. The T cells fire repeatedly during a maintained depolarization; N cells fire spontaneously and have a large undershoot. Current injected is monitored on the upper trace and the calibration is 5 × 10^{-9} A. (Modified from Nicholls and Baylor 1968.)

skin surface. Brief indentation of 30 μm, or even eddys in the solution bathing the skin, cause the neurons to respond. The sensory discharge of a T cell is rapidly adapting to a step indentation and usually ceases within a fraction of a second. As might be expected, a tactile stimulus moved back and forth over the skin gives rise to a maintained discharge whose frequency can be graded by varying the rate at which the stimulus moves or by varying the rate at which the skin is indented at a point. The P cells respond only to a marked deformation

FIGURE 2

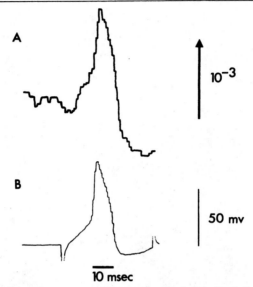

▲ Simultaneous measurement of fluorescence intensity (*A*) and membrane potential (*B*) in an N cell of a leech segmental ganglion by use of a merocyanin dye. (↑) The stated value of the change in intensity, *I*, divided by the resting intensity, I_r, in a single sweep. The response time constant of the light recording system was 2.2 msec. The signal averager used for recording sampled once every 0.5 msec and this accounts for the jagged character of the records. Cell body diameter, 68 μm; temperature, 21°C. (Reprinted, with permission, from Salzberg et al. 1973.)

of the skin. Their discharge is slowly adapting and lasts 20 sec or more during maintained pressure. Again, the frequency is graded with the extent of the indentation. Light touch is ineffective in activating P cells. The N cells require still stronger mechanical stimuli. The threshold has not been determined precisely, but the stimulus on the skin that gives the highest frequency and best-maintained discharge is a radical deformation produced by pinching the skin with forceps or scratching it with a pin. The N cells, like the P cells, are slowly adapting and often continue to fire after the stimulus has been removed. Similar T, P, and N cells can be readily identified in *Haementeria*, except that each ganglion contains a single N cell on either side.

The three kinds of sensory cells — totaling 14 per ganglion — respond specifically and selectively to mechanical stimuli and are not activated by changes in the temperature, acidity, or osmotic pressure

FIGURE 3

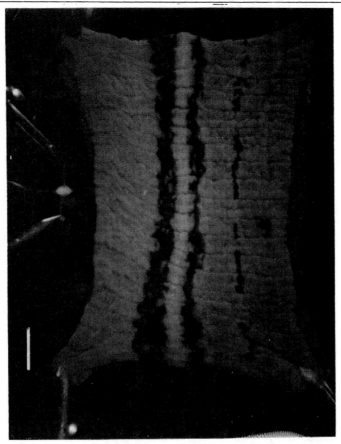

▲ Part of the body wall of the leech pinned out with a segmental ganglion attached by its roots. The skin has been taken from the dorsal midline to the ventral midline. Dorsal skin is dark and is marked by orange stripes. Lateral skin is marked by a large black stripe. Annular margins run circumferentially. Scale, 2 mm. (Reprinted, with permission, from Nicholls and Baylor 1968.)

of the bathing fluid. There is a striking parallel between these mechanosensory cells in the leech and those that innervate the human skin, which also distinguish between touch, depression, and noxious or "painful" stimuli. Even before the anatomical evidence was available (see below), a number of physiological experiments indicated that these cells in the leech are true sensory cells, rather than second- or third-order neurons driven only indirectly by sensory cells in the

FIGURE 4

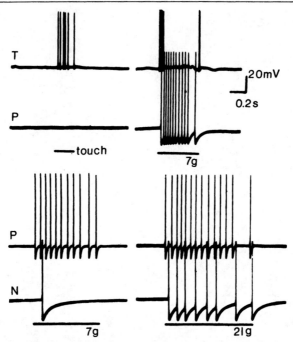

▲ Intracellular recordings from T, P, and N cells to illustrate their responses to cutaneous mechanical stimuli (see Fig. 3 for photograph of the preparation). (*Upper traces*) Simultaneous recordings from T and P cells during natural stimulation. (*Left*) Touching the skin lightly (0.5g) caused the T cell to fire, whereas the P cell remained silent. (*Right*) The skin was indented by a 200-µm diameter stylus attached to a 7g spring; this stimulus caused a maintained discharge of the P cell and rapidly adapting discharges of the T cell upon application and release of the stimulus. (*Lower traces*) Simultaneous records from P and N cells. (*Left*) The 7g stimulus fired only the P cell (the single N-cell action potential occurred "spontaneously" as part of a regular rhythm of firing). (*Right*) A 21g stimulus applied in the same way fired both cells. Note that P and N cells adapt slowly, whereas the T cell adapts rapidly. (Reprinted, with permission, from Nicholls and Baylor 1968.)

periphery, and that they are the principal cells conveying sensory information about touch, pressure, and noxious mechanical stimuli to the segmental ganglia. Each of the 14 sensory cells innervates a clearly defined area of skin and responds only to stimuli applied within one of these circumscribed receptive fields. The boundaries of a field can be identified by landmarks, such as segmentation or the coloring of skin, so that one can predict reliably which cells will fire when a particular area is touched, pressed, or pinched. Comparable experiments have been made by A.E. Stuart and by G. Stent and his colleagues C. Ort and W. Kristan (Stuart 1970; Kristan et al. 1974) to identify the motor cells and their fields of innervation.

The leech performs only a limited repertoire of simple movements. These include shortening the body in response to cutaneous stimuli, swimming, twisting, and walking like an inchworm by using the suckers. The main muscles that execute these movements are arranged in three layers. Directly under the skin lie circular fibers, which, by contracting, produce elongation of the animal. At a deeper level lie criss-crossed oblique fibers, and deeper still are powerful longitudinal fibers used for shortening and bending. There are, in addition, dorsoventral muscles that flatten the animal and others that erect the annuli into ridges.

The muscles of the leech are controlled by excitatory and inhibitory motor neurons present in the segmental ganglia. Each motor neuron innervates a group of muscle fibers that has a consistent size and location from segment to segment. To determine which cells are the motor neurons, cells are stimulated through intracellular electrodes while one observes whether any part of the body wall contracts. If contractions are observed, a number of additional tests must then be made to confirm that the axon of the stimulated cell synapses directly on the muscle fibers. Stimulation of many cells that are not motor neurons can induce contractions indirectly, by activating motor cells. One unequivocal demonstration of the role of a motor neuron is provided by the use of protease (see Appendix A). After killing the AE motor neuron by protease injection, the annuli that it alone innervates can no longer be induced to erect.

MECHANOSENSORY CELLS

Morphology of Mechanosensory Neurons within the CNS

The shapes of T, P, and N cells within the ganglion have been studied by intracellular injection of horseradish peroxidase (HRP) (Muller and McMahan 1976; Muller 1979; see Chapter 6, Figs 2 and 4). All have some basic morphological features in common — a single process arises from the cell body and large primary branches pass through the neuropil and leave the ganglion in peripheral nerve roots. However, the shape, length, and distribution of the numerous secondary processes that arise from the main axon are different and distinctive for each of the three modalities of neuron (see Fig. 2 in Chapter 6), as are

FIGURE 5

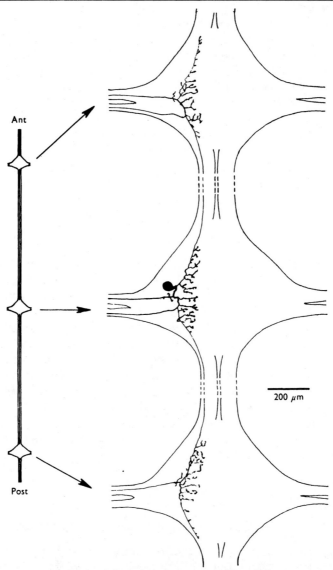

▲ Camera lucida drawing of an injected T cell that innervated ventral skin. The entire extent of the arborization of the cell spanned three segmental ganglia. The processes that left the CNS through the roots of the anterior and posterior adjacent ganglia were more slender than those that went through the roots of the cell's own ganglion. Note the asymmetry in branching in the anterior and posterior directions. (Reprinted, with permission, from Yau 1976b.)

the shapes and arrangements of the varicosities where the synapses are made. T cells have the simplest branching pattern, with short secondary branches that do not cross the midline of the ganglion; N cells have the most elaborate, with several fine secondary branches extending through the neuropil on both sides of the ganglion. In *Haementeria*, the branching patterns are similar but less complex (Kramer and Goldman 1981).

Yau showed that in addition to synapsing in their own ganglion, the mechanosensory cells have processes in the connectives to neighboring ganglia, where they also synapse and send branches to the periphery (Fig. 5; Yau 1976b). The arborization of a mechanosensory neuron in a midbody segment thus spans three segmental ganglia. In ganglia near the head or tail ends of the chain, the arborization may span more than three ganglia. Thus, a T cell near the head arborizes in a second, more rostral ganglion; a T cell near the tail arborizes in several more caudal ganglia. Wallace (1980) has shown that motor, as well as sensory, cells arborize more extensively in ganglia at the head or tail ends of the leech. There is also evidence that neurons in embryonic or juvenile leeches arborize over more ganglia than their homologs in the adult, raising the possibility that certain processes of a neuron disappear during embryogenesis. The ability to change shape is not confined to embryonic neurons. Following damage to the CNS, adult mechanosensory cells have been shown to sprout both centrally (Chapter 10) and peripherally (Van Essen and Jansen 1977, see below).

Receptive Field Organization

The striking feature of the innervation of the skin by mechanosensory cells is its orderliness. In each of the 21 segmental ganglia along the length of a leech, individual T, P, or N cells innervate the same specific areas of skin. For example, there are six T sensory cells in each ganglion symmetrically arranged in pairs, three on either side of the ganglion. The most laterally placed T cell innervates a discrete patch of dorsal skin on one side of the leech, the middle cell innervates ventral skin, and the medial cell innervates lateral skin (Fig. 6). Physiological mapping reveals that within the receptive field there are discrete spots sensitive to touch. All such points have the same sensitivity, but the density is highest in the central part, tapering off towards the periphery.

The specificity of skin innervation is carried further in that each cutaneous branch of a T-cell axon innervates a specific part of that cell's field (Nicholls and Baylor 1968; Yau 1976b; Kramer and Goldman 1981). The large-caliber axon branches in the anterior and posterior roots of the ganglion supply major fields in that body segment. The finer-caliber axon branches in the anterior and posterior connectives that arborize in adjacent ganglia innervate accessory fields in neighboring body segments. The boundaries between the subfields are sharp and usually correspond to the edge of an annulus (Fig. 7). Thus, the innervation pattern of a single neuron and its numerous branches appears like a patchwork quilt, each branch innervating a discrete territory in the skin with no overlap by other branches of the same cell.

Although different branches of the same axon do not overlap, innervation of a patch of skin by a T cell does not preclude the

FIGURE 6

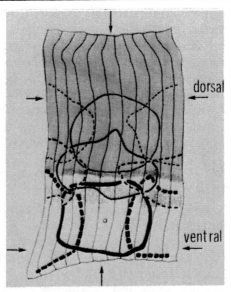

▲ Receptive fields of T cells drawn on a tracing of the skin. (→) The dorsal and ventral midlines of the skin; (↓) the central annulus, in which the sensillum appears as a small circle; (⎯⎯⎯, ⎯⎯⎯) boundaries of the receptive fields of the three T cells; (— — — —, ---------) the fields of cells in adjacent ganglia. Fields were mapped successively for the nine cells using light touch and marking the positions from which responses could be obtained on an enlarged photograph of the skin. Cells could be driven most effectively by touching the center of their fields; fewer action potentials were obtained by stimulation at the edges of the fields. An annulus width is approximately 1 mm. (Reprinted, with permission, from Nicholls and Baylor 1968.)

presence of other T cells, or of P or N cells. Because each T cell innervates skin in three body segments, there is extensive overlap along the length of a leech by the receptive fields of T cells in neighboring ganglia (Fig. 8). There is also a small amount of overlap dorsoventrally between the fields of T cells within a ganglion.

In the subesophageal ganglion in the head of *Hirudo*, neurons have been identified that are homologous to the T, P, and N cells in the segmental ganglia, with similar electrical properties and responses to skin stimulation (Yau 1976a). In the head region, the regular arrangement of body segments with five annuli is disrupted by loss of annuli from particular segments and by the presence of the mouth and anterior sucker on the ventral aspect. The head mechanosensory neurons innervate territories on the external surface of the head and on the oral folds of the anterior sucker. The territories overlap, as do the segmental receptive fields, and in some cases they are completely superposed.

That a number of cells of the same modality can innervate a small patch of skin has also been demonstrated in an abnormal leech in which, by developmental chance, each segmental ganglion had more than its usual complement of sensory and motor cells. Kuffler and Muller (1974) showed that the extra T, P, and N cells resembled those in normal ganglia with respect to their sizes, positions, electrical properties, modalities, and branching patterns of their axons in the various nerve roots. Mapping of the receptive fields of the supernumerary cells showed that two neurons of one modality could independently innervate overlying regions of skin that normally would be supplied by one cell. Thus, skin can receive innervation over and above its normal quota.

Reinnervation of Skin and Spread of Receptive Fields after Deletion of Single Cells

After cutting or crushing peripheral nerve roots, the mechanosensory axons will regenerate to reinnervate the skin (Van Essen and Jansen 1977). Moreover, the repair usually takes place with a high degree of precision. Invariably, the cells regain their appropriate modality, and although some cells show clear abnormalities in the extent or position of their receptive fields after regeneration, several T cells completely reestablish their peripheral fields so that the pattern of innervation after regeneration resembles that originally laid down during development.

These features of receptive field organization raise a number of questions as to how the innervation is established. For example, do

FIGURE 7

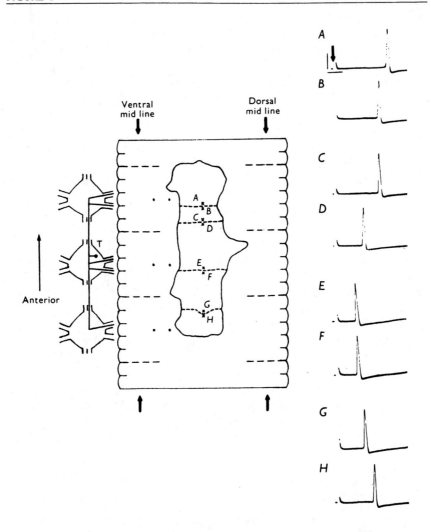

▲ Subfields of a T cell that innervated lateral skin. Each subfield was innervated by a separate branch of the axon passing through either a root of its ganglion or that of an adjacent ganglion. Adjacent subfields had negligible overlap with each other, as indicated by discrete jumps in the time delay of intracellularly recorded action potentials (*right*) when a mechanical stimulus was moved across the boundary between two adjoining subfields. Vertical calibration, 20 mV; horizontal calibration, 10 msec. (↓) The time when the mechanical stimulus was applied. (Reprinted, with permission, from Yau 1976b.)

two neighboring cells of the same modality compete for territory in the skin. And, if so, can the receptive field of a neuron spread if the surrounding cells are removed without damage to the neuron itself? S.E. Blackshaw, J.G. Nicholls, and I. Parnas (in prep.) have shown that if three out of the four N cells in a segmental ganglion are killed by protease injection (see Appendix A) and the leech is subsequently allowed to recover, the receptive field of the remaining N cell expands to take over the denervated territory on the contralateral side of the leech, a region it does not normally innervate (Fig. 9). Similarly, the field of the T cell that innervates dorsal skin spreads across the midline to innervate contralateral skin after the three T cells on that side have been deleted (Fig. 10).

Morphology of Sensory Terminals of T Cells

To approach problems of how innervation is established, it is necessary to have information about (1) the morphology of the neurons in the skin, (2) the structure of the sensory terminals, (3) their distribution in the receptive field, and (4) their structural relationship with peripheral cells. By injecting HRP into the cell bodies of individual neurons in the ganglion and allowing it to spread to the skin, it is possible to see the terminals in whole mounts of the body wall and to construct maps of the number of terminals and their distribution within the cell's receptive field (Blackshaw 1981). In this way, T-cell axons are found to branch where the nerve roots divide in the body wall and to become progressively finer as they dip between the layers of body wall muscle. When they reach the layer of epithelial cells in the skin, they branch extensively, forming beaded chains that turn between the epithelial cells to end 1–2 μm from the skin surface in intercellular spaces immediately below the junctional complex at the outer ends of the epithelial cells (Fig. 11). The varicosities contain a single large mitochondrion, and, in addition, the terminal varicosity contains a cluster of 30-nm vesicles. The distance at the skin surface between neighboring terminals of the same axon (between 15 μm and 150 μm) fits well with the distribution predicted in the earlier physiological mapping of T-cell receptive fields. Counts of the number of terminals of one axon branch show that a T cell makes about 200 terminals within 1 mm^2 of skin in the center of its territory (Fig. 12). Because T-cell receptive fields extend longitudinally over three body segments, covering an area of 10–20 mm^2 of skin, a single neuron is estimated to have several hundred terminals at the skin surface. Physiological mapping has shown that each of these terminals can

FIGURE 8

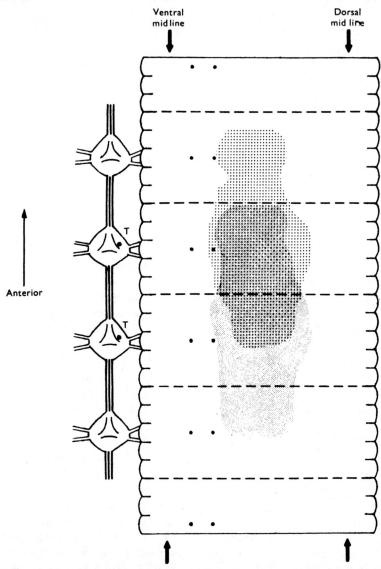

▲ Receptive fields of two T cells in adjacent ganglia that both innervate lateral skin. Each receptive field spans 12–13 annuli over three body segments, and, consequently, there is extensive overlap between the receptive fields. The size and appearance of T-cell receptive fields on ventral and dorsal skin are very comparable. (Reprinted, with permission, from Yau 1976b).

independently transduce a mechanical stimulus to the skin. This raises the problem of how the position of a stimulus is discriminated in the leech.

P- and N-cell Receptive Fields and Terminals

In *Hirudo* and in other leeches such as *Haementeria*, each segmental ganglion contains two pairs of P mechanosensory neurons (Nicholls and Baylor 1968; Kramer and Goldman 1981). The receptive field boundaries of P cells are different from those of T cells; each P cell innervates a dorsal or ventral quadrant of the segment. Thus, the P cell situated laterally in the ganglion innervates ventral skin and the medially situated P cell innervates dorsal skin by way of the dorsal branch. HRP injection shows that P cells branch far less extensively within their receptive fields than T cells do and make far fewer terminals. The axon branches appear to end deep within the skin epithelial layer, although the morphology of their terminals has not been studied in detail.

Mapping of N-cell territories in *Hirudo* shows that their receptive fields are more extensive than was previously thought (S.E. Blackshaw et al., in prep.). Both medial and lateral N neurons innervate an area of skin that extends from the ventral midline to the dorsal midline, and thus the fields of the two cells on each side are approximately coincident. Tracing the arborization of the axons in the body wall by HRP injection shows that the large axon branches in the main nerve roots give rise to fine-caliber branches about 1 μm in diameter that run within the network of peripheral nerves at deep levels of the body wall. Some of these fine branches can be traced to superficial layers of the body wall, where they run at the base of the layer of epithelial cells in the skin; for example, fine branches encircle the nephridiopore in each body segment. These superficial axon branches appear to end below the skin epithelial layer and have never been seen to terminate between epithelial cells at the surface like T cells do. In addition to innervating more superficial layers, the lateral N cell makes terminals with a distinctive coiled appearance at deeper levels of the body wall in association with large neurons (HOs or Hoovers), whose cell bodies lie within the sheath of a peripheral nerve (Fig. 11). These peripheral neurons are distinct from the monoamine-containing cells in peripheral nerves that were described by Rude (1969) and have unknown function. They are characterized by fan-shaped dendrites associated with the longitudinal muscles of the body wall; this morphology suggests that they are stretch receptors. The coiled terminals of the N

FIGURE 9

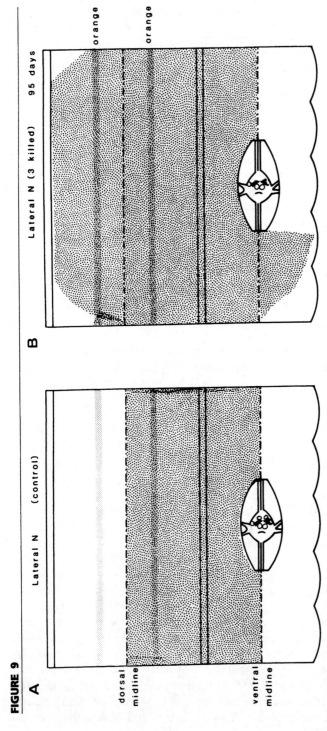

▲ Normal and expanded receptive fields of N sensory cells. (A) Schematic representation showing that each of the two N cells has a receptive field that extends from the dorsal midline to the ventral midline in normal animals. (B). Expanded receptive field of a lateral N cell in a ganglion in which the three other N cells had been killed 95 days beforehand. The cell innervated almost the entire territory, including contralateral skin up to the edge, an area it does not normally supply. Some spread occurred across the ventral midline, but this region was inevitably damaged during dissection. Thresholds for electrical and mechanical activation were similar in the normal and expanded fields. The fields of T and P cells in this preparation did not cross the dorsal midline.

FIGURE 10

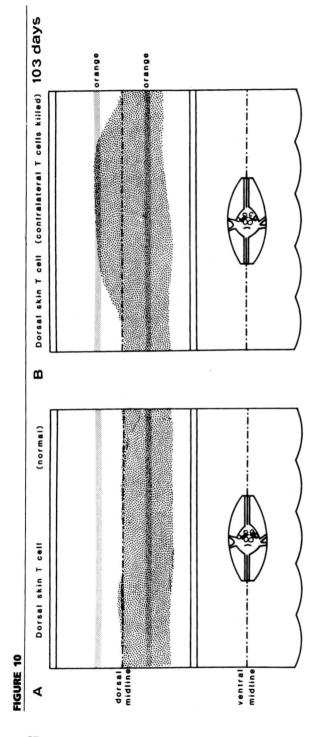

▲ Expansion of T-cell fields following deletion of three contralateral T cells. (A) Normal receptive field of T cell innervating dorsal skin. (B) 103 days after deletion of contralateral T cells, the field of the dorsal T cell had expanded across the dorsal midline to reach the contralateral orange stripe. Sensitivity to light touch was similar in normal and expanded territories. No expansion of N- or P-cell fields occurred.

cells arborize homodromically over the dendrites of the HO cells (hence the name).

The finding that N cells contact these peripheral neurons raises the

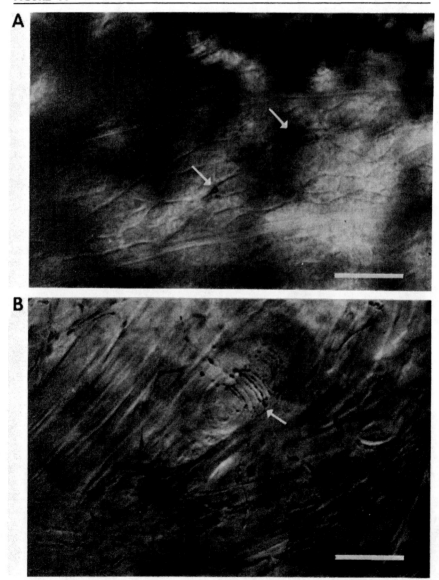

FIGURE 11

Sensory Cells and Motor Neurons

possibility that N cells respond to some other form of stimulation in addition to noxious stimulation of the skin, perhaps stretch of the body wall. This has led to a recent reexamination of their role that, however, clearly confirms the earlier work. N cells do not respond to light touch, steady pressure, or moderate changes in osmolarity, pH, or temperature, or to stretching the skin or underlying muscles. The problem as to the exact role of the N-cell terminal on the peripheral neurons therefore remains.

Interestingly, the medial N cell, but not the lateral, in addition to responding to noxious mechanical stimulation of the skin, also responds to similar stimuli applied to the connective tissue around the gut (Fig. 13). Furthermore, it is the lateral N cell, and not the medial, that has the distinctive coiled terminals on peripheral neurons. Pharmacological differences in extrasynaptic receptors between the medial and lateral N cells have also been reported — the lateral cell is depolarized by acetylcholine (ACh), 5-hydroxytryptamine (5-HT), γ-aminobutyric acid (GABA), and glycine; the medial cell is depolarized only by ACh and hyperpolarized by dopamine and noradrenaline (Sargent 1977). Thus, although lateral and medial N cells have the same modality and identical electrical properties and have been shown by Zipser and McKay (1981) to share immunologically unique macromolecules (see Appendix B), there is increasing evidence of a number of differences between the two neurons.

Signaling of Sensory Information by Sensory Cells

Two important considerations for signaling are: (1) the arrangement of overlapping fields and (2) conduction block in the branching axons. Because the fields of T cells in neighboring ganglia overlap extensively, lightly touching a patch of dorsal skin, for example, may activate the terminals of three dorsal T cells. Impulses will enter the CNS in the large-caliber axon branches of the T cell in that segment and also in the fine accessory branches of the T cells whose cell bodies lie in the

◄ Mechanosensory cell terminals visualized in whole mounts of the body wall after injection of HRP into their cell bodies in the ganglion. *(A)* Nomarski photograph of the surface of ventral skin showing the profiles of the skin epithelial cells and two adjacent terminals (↓) of the T cell innervating ventral skin lying between the epithelial cells at the skin surface. *(B)* Nomarski picture showing a coiled terminal (↓) of the lateral N cell in association with a large neuron (HO cell) whose cell body lies within the sheath of a peripheral nerve. Scale bar, 25 μm.

FIGURE 12

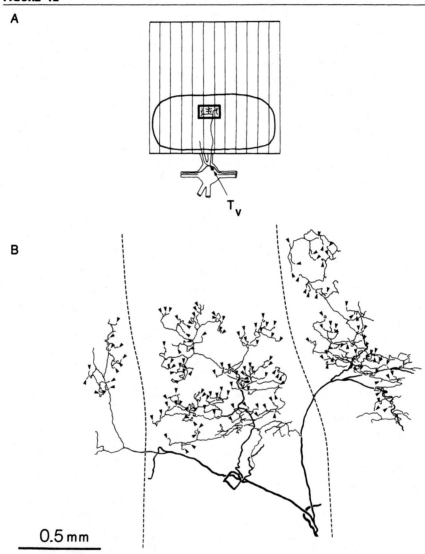

▲ Distribution of T-cell terminals in ventral skin. *(A)* Schematic diagram showing the cell body of the T cell, T_v, in the ganglion, and its receptive field (oval) in the ventral surface of the leech. *(B)* Camera lucida drawing of the aborization and endings (▲) of the neuron within 1 mm² of skin from the center of the cell's receptive field contained in the rectangle in *A*. The T cell made 178 endings within this central part of its receptive field; 108 endings were located on the central annulus, and 70 were on part of the annuli on either side. (Reprinted, with permission, from Blackshaw 1981.)

adjacent ganglia (see Fig. 8). Each of these three T cells makes synapses in three adjacent ganglia. Thus, light touch on a patch of skin can activate T-cell synapses in as many as five neighboring ganglia. Similar considerations apply to P and N cells.

How far impulses travel in a neuron depends also on its previous activity. Baylor and Nicholls (1969) showed that trains of impulses in mechanosensory cells, such as occur naturally during intense stimulation of the skin, lead to a long-lasting hyperpolarization of the membrane. The prolonged discharge of a T cell, for example, may be followed by a hyperpolarization of up to 30 mV, which declines gradually over several minutes (Fig. 14). Two factors were shown to contribute to the hyperpolarization of mechanosensory neurons after activity: (1) the activity of an electrogenic Na^+ pump and (2) a prolonged increase in K^+ conductance whose amplitude and duration depends on the external concentration of Ca^{++}. In N cells, the increase in K^+ conductance is the principal factor responsible for the hyperpolarization; in T cells it is primarily activation of the Na^+-K^+ pump; whereas in P cells both are important (Jansen and Nicholls 1973). As a result of the hyperpolarization, the threshold to mechanical stimuli is raised wherever impulses have traveled. Repetitive firing may also lead to conduction block in branches of the T or P cell (Van Essen 1973; Yau 1976b; Muller and Scott 1981; see also Chapter 6). If conduction block occurs where fine axons join larger axons in the periphery, then impulses fail to reach the CNS. Yau (1976b) showed that conduction

FIGURE 13

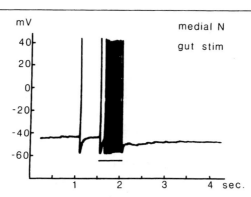

▲ Response of the medial N cell to noxious stimuli applied internally. Gentle pinching or stretching of the internal layers of connective tissue lining the gut elicited a high-frequency discharge in the N cell.

FIGURE 14

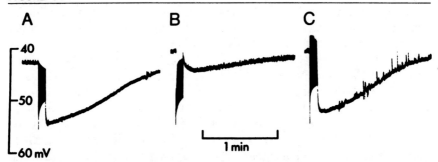

▲ Effect of strophanthidin on the hyperpolarization of a pressure neuron elicited by electrical stimulation of its axon. Strophanthidin blocks the Na$^+$-K$^+$-dependent ATPase. The electrode remained in the cell while the composition of the bathing fluid flowing past the preparation was changed. *(A)* Control response to standard train at 40/sec for 6 sec. *(B)*. Same, 2 min after applying 1.3 × 10^{-4} M strophanthidin. *(C)* Recovery of response after 15-min wash in normal Ringer's fluid. (Reprinted, with permission, from Baylor and Nicholls 1969.)

block may also occur at branching points within the CNS, most probably where fine axons from the accessory fields join the main axon. T and P cells make synapses with other cells in three ganglia (see Fig. 5), and impulses that become blocked where the axon branches within the CNS may therefore activate some synapses independently of others (Muller and Scott 1981). The geometry of the neuron is therefore of paramount importance for the transmission of information to higher-order cells.

Other Sensory Receptors

Studies on sensory neurons in the leech have been concerned mainly with mechanosensory neurons that innervate the skin, because of their large size, accessibility, and ease of identification. Of other modalities of mechanosensory neurons, the group of large peripheral HO cells in every segment that are innervated by the N cell appear on morphological grounds to be stretch receptors, although their role has yet to be confirmed physiologically. Stretch receptors have not previously been described in the leech, although they have been postulated on theoretical grounds by Kristan and Stent (1976) as being part of the proprioceptive feedback to the rhythmically active swimming motor neurons. It has been known since the 19th century from the drawings of Methylene Blue-stained ganglia made by Retzius that large unidentified axons enter the ganglion from the periphery, and it seems likely that these are the centrally directed processes of the HO cells.

Hirudo possesses two types of photosensory organ: the 10 paired eyes on the dorsal surface of the head and the 14 sensilla that lie around the middle annulus of every body segment. Both the eyes and the sensilla contain a small number of spherical refractile cells whose axons project to the CNS and which mediate the response to light (Kretz et al. 1976). Phillips and Friesen (1981) have shown that the sensilla also contain, in addition to the photoreceptors, three different types of ciliated cells, one of which is thought to be a primary sensory neuron responsible for the detection of water movements (Fig. 15). Activation of the receptors leads to a volley of small action potentials in the root and the occurrence of IPSPs in the AE motor cell.

MOTOR NEURONS

Motor neurons in the leech were first identified and characterized by Ann Stuart (1970). She described 34 neurons that could be recognized consistently in each of the 21 segmental ganglia. These neurons comprise 14 pairs of excitatory motor neurons and 3 pairs of inhibitory motor neurons; together they supply the five groups of muscles in each segment responsible for the movements of a leech.

Thirty of the 34 neurons are located on the dorsal surface of the ganglion and innervate the three layers of muscle that make up the major part of the body wall. These are the circular, oblique, and longitudinal muscle fibers that form continuous sheets along the length of a leech. They themselves have no obvious segmental boundaries but they are innervated segmentally by the 21 ganglia. Of the remaining two pairs of neurons, the AE cells on the ventral surface of the ganglion innervate the fourth group of muscle fibers. These insert into skin on either side of the annulus, and their contraction produces erection of the annuli (Fig. 16). This reflex is elicited by mechanical stimulation of the skin and is mediated by a monosynaptic pathway from the mechanosensory neurons to the AE motor neurons, although its significance to the leech is rather obscure. A pair of flattener motor neurons innervates the fifth group of muscle fibers which run from the dorsal to the ventral surface and which, when contracted, flatten the leech. A number of motor neurons, principally those innervating the dorsoventral muscles and the longitudinal muscle layer, are involved in swimming in the leech. These and additional motor neurons to the longitudinal muscles have been further characterized by Ort et al. (1974) and others (see Appendix D). Other motor neurons that have

FIGURE 15

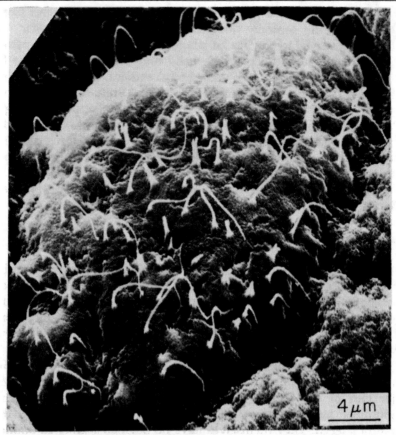

▲ Scanning electron micrograph of the central area of a sensillum showing two distinct types of filiform processes: single S hairs and grouped G hairs. The S hairs may be the sensory structures mediating sensitivity to low-amplitude water movements. (Photograph kindly provided by Dr. W. O. Friesen.)

been functionally identified are the paired heart excitor (HE) cells on the ventral surface of each ganglion, which segmentally innervate the muscles of the heart tube, and motor neurons in the "sex" ganglia of the leech, which supply the reproductive organs and which were identified by Zipser (1979a,b).

The branching patterns of motor cells revealed by dye injection differ markedly from those of sensory cells. Most, but not all, send a process across the ganglion to innervate muscles via contralateral nerve roots (see Chapter 6; Fig. 4). They differ particularly in the arrangement of the numerous highly branched and densely packed secondary branches that radiate from the main process and its secondary branches (see, e.g., Muller and McMahan 1976). In contrast to mechanosensory cells, overshooting action potentials are not recorded from motor neuron cell bodies, presumably because impulses are initiated in the axons as they leave the ganglion and do not invade the cell body or branches within the ganglion (but see Fuchs et al. 1981).

Fields of Innervation by Motor Neurons

The territories innervated by individual motor neurons have been mapped by watching which muscles contract when a neuron is stimulated intracellularly. A general principle that is probably of develop-

FIGURE 16

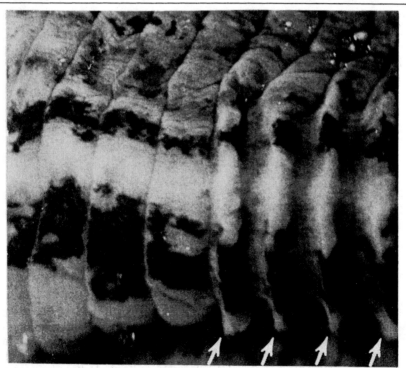

▲ Erection of annuli. The AE motor neurons on the ventral surface of the ganglion innervate muscle fibers which insert into skin on either side of an annulus. Contraction of the muscles produces erection of the annuli (↓).

mental significance is that cells that innervate neighboring territories within the same muscle layer tend to lie together within the ganglion, their fields overlapping to a small extent. For example, Stuart (1970) identified six pairs of excitatory motor neurons to the longitudinal muscle layer. This is the thickest of the three layers of body wall muscle and consists of a number of discrete bands of fibers separated by connective tissue. Five of the six pairs of longitudinal motor neurons innervate particular groupings of fibers (Fig. 17). Three neurons lie together next to the anterior connective and divide the muscle into dorsal, lateral, and ventral strips; the other two neurons lie adjacent in a posterior group of cells and divide the longitudinal muscle into two strips, dorsal and ventral, on either side of the lateral black stripe. The sixth pair of excitatory motor neurons to the longitudinal muscles, lying alone, is the pair of large longitudinal motor neurons (L cells), each of which innervates all the longitudinal muscle fibers on one side of the segment. The two cells in a ganglion are coupled by an electrical synapse that synchronizes their activity, ensuring symmetrical contraction of the entire body segment. A shortening reflex occurs in response to noxious stimuli and is mediated by a monosynaptic pathway from the mechanosensory neurons to the L cell (Chapter 6).

Each motor neuron innervates a field that extends longitudinally

FIGURE 17

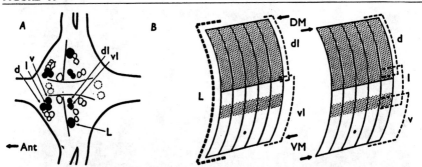

▲ *(A)* Positions in the ganglion of excitatory cells supplying longitudinal musculature. Dorsal surface of the ganglion. *(B)* Approximate circumferential extent of the territories of these cells as related to pigment markings on a piece of skin. DM = dorsal midline; VM = ventral midline. Cells are labeled in *A* with small letters indicating the position of their innervated territory, as seen in *B*: d = dorsal, l = lateral, v = ventral, dl = dorsolateral, vl = ventrolateral. The cell labeled L innervates fibers from one midline to the other. (Reprinted, with permission, from Stuart 1970.)

into adjacent body segments and there overlaps with the fields of homologous neurons. Stuart (1970) made a detailed map of the territory of the AE motor neuron and showed that it had a quiltlike pattern, resembling receptive fields of T mechanosensory cells, with individual branches innervating specific parts of the cell's field and sharp boundaries between the subfields.

As with sensory neurons, Van Essen and Jansen (1977) showed that motor neurons can grow back to reinnervate their territory in the periphery after the roots have been cut or crushed. The L motor neuron reinnervates longitudinal muscles and the AE motor neuron grows through longitudinal muscles and reinnervates AE muscles with a high degree of precision.

REFERENCES

Baylor, D.A. and J.G. Nicholls. 1969. After-effects of nerve impulses on signalling in the central nervous system of the leech. *J. Physiol.* **203**:571–589.

Blackshaw, S.E. 1981. Morphology and distribution of touch cell terminals in the skin of the leech. *J. Physiol.* (in press).

Fuchs, P.A., J.G. Nicholls, and D.F. Ready. 1981. Membrane properties and selective connexions of identified leech neurons in culture. *J. Physiol.* (in press).

Grinvald, A., L.B. Cohen, S. Lesher, and M.B. Boyle. 1981. Simultaneous optical monitoring of activity of many neurons in invertebrate ganglia using a 124-element photodiode array. *J. Neurophysiol.* **45**:829–840.

Jansen, J.K.S. and J.G. Nicholls. 1973. Conductance changes, an electrogenic pump and the hyperpolarization of leech neurones following impulses. *J. Physiol.* **229**:635–665.

Kramer, A.P. and J.R. Goldman. 1981. The nervous system of the glossiphonid leech *Haementeria ghilianii*. I. Identification of neurons. *J. Comp. Physiol.* (in press).

Kretz, J.R., G.S. Stent, and W.B. Kristan, Jr. 1976. Photosensory input pathways in the medicinal leech. *J. Comp. Physiol.* **106**:1–37.

Kristan, W.B., Jr. and G.S. Stent. 1976. Peripheral feedback in the leech swimming rhythm. *Cold Spring Harbor Symp. Quant. Biol.* **40**:663–674.

Kristan, W.B., Jr., G.S. Stent, and C.A. Ort. 1974. Neuronal control of swimming in the medicinal leech. I. Dynamics of the swimming rhythm. *J. Comp. Physiol.* **94**:97–119.

Kuffler, D.P. and K.J. Muller. 1974. The properties and connections of supernumerary sensory and motor nerve cells in the central nervous system of an abnormal leech. *J. Neurobiol.* **5**:331–348.

Muller, K.J. 1979. Synapses between neurones in the central nervous system of the leech. *Biol. Rev.* **54**:99–134.

Muller, K.J. and U.J. McMahan. 1976. The shapes of sensory and motor neurons and the distribution of their synapses in the ganglia of the leech: A study using intracellular injection of horseradish peroxidase. *Proc. R. Soc. Lond. B* **194**:481–499.

Muller, K.J. and S.A. Scott. 1981. Transmission at a "direct" electrical connexion mediated by an interneurone in the leech. *J. Physiol.* **311**:565–584.

Nicholls, J.G. and D.A. Baylor. 1968. Specific modalities and receptive fields of sensory neurons in the C.N.S. of the leech. *J. Neurophysiol.* **31**:740–756.

Ort, C.A., W.B. Kristan, Jr., and G.S. Stent. 1974. Neuronal control of swimming in the medicinal leech. II. Identification and connections of motor neurons. *J. Comp. Physiol.* **94**:121–154.

Phillips, C.E. and W.O. Friesen. 1981. The fine structure of leech sensilla. *Neurosci. Abstr.* **7**:00. (in press).

Retzius, G. 1891. *Biologische Untersuchungen, Neue Folge II.* Sampson & Wallin, Stockholm

Rude, S. 1969. Monoamine-containing neurons in the central nervous system and peripheral nerves of the leech, *Hirudo medicinalis. J. Comp. Neurol.* **136**:349–372.

Salzberg, B.M., H.V. Davila, and L.B. Cohen. 1973. Optical recording of impulses in individual neurones of an invertebrate central nervous system. *Nature* **246**:508–509.

Sargent, P.B. 1977. Transmitters in the leech central nervous system: Analysis of sensory and motor cells. *J. Neurophysiol.* **40**:453–460.

Sargent, P.B., K.-W. Yau, and J.G. Nicholls. 1977. Extrasynaptic receptors on cell bodies of neurons in central nervous system of the leech. *J. Neurophysiol.* **40**:446–452.

Stuart, A.E. 1970. Physiological and morphological properties of motoneurones in the central nervous system of the leech. *J. Physiol.* **209**:627–646.

Van Essen, D.C. 1973. The contribution of membrane hyperpolarization to adaptation and conduction block in sensory neurones of the leech. *J. Physiol.* **230**:509–534.

Van Essen, D.C. and J.K.S. Jansen. 1977. The specificity of re-innervation by identified sensory and motor neurons in the leech. *J. Comp. Neurol.* **171**:433–454.

Wallace, B.G. 1980. Selective neurite atrophy during development of cells in the leech C.N.S. *Neurosci. Abstr.* **6**:679.

Yau, K.-W. 1976a. Physiological properties and receptive fields of mechanosensory neurones in the head ganglion of the leech: Comparison with homologous cells in segmental ganglia. *J. Physiol.* **263**:489–512.

———. 1976b. Receptive fields, geometry and conduction block of sensory neurones in the C.N.S. of the leech. *J. Physiol.* **263**:513–538.

Zipser, B. 1979a. Identifiable neurons controlling penile eversion in the leech. *J. Neurophysiol.* **42**:455–464.

———. 1979b. Voltage-modulated membrane resistance in coupled leech neurons. *J. Neurophysiol.* **42**:465–475.

Zipser, B. and R. McKay. 1981. Monoclonal antibodies distinguish identifiable neurones in the leech. *Nature* **289**:549–554.

6
Synapses and Synaptic Transmission

Kenneth J. Muller
Department of Embryology
Carnegie Institution of Washington
Baltimore, Maryland 21210

Synapses in the nervous system of the leech are strikingly similar in their morphology and in their functional properties to those found in higher animals. Thus, synapses communicating via chemical transmitters (chemical synapses) and via electrical junctions (electrical synapses) both occur in the leech nervous system. It has been possible to focus on the ways in which individual, functionally identified leech neurons process and selectively transmit information to other neurons. Not only are the shapes, positions, and functions of identified leech neurons consistent from ganglion to ganglion or from animal to animal, but so are their synaptic connections. This high degree of consistency makes it possible to study how short-term and long-term changes come about at synaptic connections between specific cells.

In their study of the morphology of leech synapses, Purves and McMahan (1972) were able to recognize synaptic terminals on functionally identified leech motor neurons by injecting these cells with the dye Procion Yellow. Another intracellular marker that has been injected and one that is particularly suitable for electron microscopy as well as light microscopy is horseradish peroxidase (HRP) (Muller and McMahan 1975). By use of HRP as a marker, synapses of identified cells have been characterized extensively, and there has been a rapid convergence of morphology and physiology in the analysis of the neural network. This chapter presents an overview of the varieties of specific synapses and of their function, organization, and distribution in the leech nervous system.

SYNAPSE STRUCTURE

Chemical Synapses

Under the electron microscope, particular points of contact between leech neurons have been identified as chemical synapses on the basis

FIGURE 1

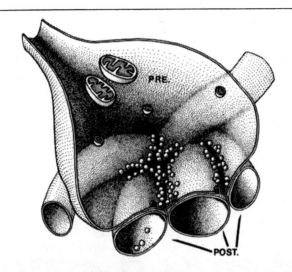

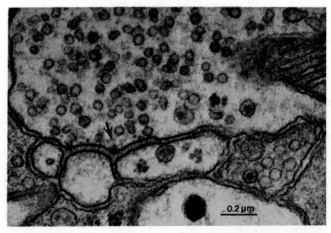

of their morphological resemblance to well-characterized neuromuscular junctions (Couteaux and Pecot-Dechavassine 1970; Dreyer et al. 1973; Heuser et al. 1974), and in a very few cases physiological and morphological techniques have also been brought to bear on the same synaptic structures between neurons (e.g., McMahan and Kuffler 1971). In their electron microscopy study of the leech nervous system, Coggeshall and Fawcett (1964) described synaptic structures that they found in the neuropil. The principal feature of these synapses, which in the first study seemed to be fairly homogeneous, is a cluster of synaptic vesicles in the presynaptic terminal apposed to the presumed postsynaptic element. Purves and McMahan (1972) refined the analysis of chemical synapses in the leech, noting that, as at chemical synapses in other nervous systems, so also in the leech pre- and postsynaptic cells are separated by a uniformly widened (~30 nm) extracellular synaptic cleft. There are small vesicles clustered near a tuft of electron-dense material applied to the presynaptic membrane, and there is a pair of adjoining postsynaptic processes. The presynaptic terminal is distinctively scalloped at the synapse, and the postsynaptic processes generally are of smaller caliber than the presynaptic terminal. A more complete, three-dimensional picture of these synaptic structures was formed by Muller and McMahan (1976), who examined serial sections (Fig. 1). It transpired that Purves and McMahan's two-dimensional picture of the synapse is seen only in sections in which the synapse is cut at right angles across the band of dense material attached to the presynaptic membrane (see Fig. 1). The three-dimensional structure resembles in many respects that of other synapses, including those in

◀ *(Top)* General scheme of synaptic specializations in the leech. Diagrammatic representation of typical synapses based on serial sections. The presynaptic structure (Pre.) contains numerous agranular vesicles about 50 nm in diameter (only a few clustered next to the plasma membrane are shown) and is separated from postsynaptic processes by an extracellular space widened to about 30 nm. A narrow (~ 50 nm wide) band of dense material is attached to the inner surface of the presynaptic membrane just opposite adjoining postsynaptic profiles (Post.). Lateral to the dense band, an occasional vesicle fuses with the presynaptic membrane. At the synaptic cleft, the postsynaptic processes are in close proximity. Postsynaptic processes can also contain vesicles, and broad areas of dense material line the inner surfaces of the postsynaptic membranes. *(Bottom)* Vesicle-filled presynaptic process opposite conjoint postsynaptic processes. Cross section through presynaptic band of dense material (↓). (See Figs. 3 and 4 for micrographs of other synapses.) (Modified from Muller and McMahan 1976.)

the human retina (Dowling 1970) and the frog neuromuscular junction. Synaptic vesicles are clustered near the dense bands that correspond to the active zone at the neuromuscular junction. Moreover, vesicles often appear to be fused with the membrane adjacent to the dense band, and the apposition of the postsynaptic processes resembles in some respects the junctional folds. There is usually discernible some densely staining material applied to the inside of the postsynaptic membrane and forming a faint line within the cleft. At the chemical synapse of the leech, the postsynaptic element may or may not contain synaptic vesicles, but even when vesicles are present, the polarity of the synapse remains unequivocal because the vesicles in the postsynaptic elements are not aggregated at the synapse. Ordinarily, the presynaptic cluster of 50-nm clear-cored (agranular) vesicles is bounded by mitochondria and 100-nm dense-cored vesicles. Such synapses also have subsequently been found in other invertebrates (Armett-Kibel et al. 1977; Wood et al. 1977).

The present chapter focuses on chemical synapses of this sort, but other types of chemical synapses probably exist in the leech. For example, one commonly sees in sections through the neuropil varicose profiles containing both ~50-nm granular vesicles and 100-nm vesicles with extremely dense, somewhat eccentric cores. No specialized synaptic structures have yet been seen in serial sections through such terminals, which resemble in many respects terminals containing monoamines in other systems (see Geffen and Livett 1971). It is possible that the contents of the vesicles are simply released in the vicinity of the postsynaptic target. Another equally rare terminal, containing agranular vesicles, is marked by a uniform cytoplasmic density that quite distinguishes it from surrounding processes. It has typical synaptic specializations, such as a uniform widening of the extracellular space between the vesicle-filled terminal and an adjacent process. A densely staining tuft associated with the plasma membrane is occasionally seen within the presynaptic terminal, but the terminals have not been well characterized in serial sections. Nevertheless, these terminals differ from those described above that contain electron-lucent vesicles by having infoldings that cause external leaflets of "presynaptic" membrane to be closely apposed at the presumed synapse. Unlike for the other synapses, no particular cell has yet been identified with these types of terminals.

Distribution of Synapses of Identified Neurons

Comparison of the branching patterns within ganglia of dozens of HRP-injected mechanosensory neurons reveals that they are morphologically distinct from motor neurons, as shown previously by Purves, and furthermore that touch (T), pressure (P), and nociceptive (N) cells can be easily distinguished from each other (Muller and McMahan 1976). Among mechanosensory neurons, T cells have the simplest branching patterns and N cells the most elaborate. The arrangement of secondary processes that emerge from the main axons is characteristic of each type, as also are the shapes and distribution of enlargements or varicosities where the synapses occur. For example, varicosities on the T cells are usually 3–8 μm in diameter and occur along and at the ends of secondary processes (Fig. 2). These are less elaborate than the clusters of varicose fingers on the longer secondary processes of the P cells. The varicosities of the N cells are probably the simplest of all, resembling in some respects those of the T cell but occurring simply as slight swellings along the more extensive and highly branched processes.

There are also characteristic variations within cell types of a single mechanosensory modality. For instance, of the three T cells with a receptive field in ventral, lateral, or dorsal skin, each has a distinct branching pattern of its processes. The ventral T cell extends numerous short, often unbranched secondary processes from the main axons into the neuropil, somewhat like teeth of a comb. The dorsal T cell, in contrast, has fewer but more highly branched processes whose total length and number and distribution of varicosities are the same as for the ventral T cell. The lateral T cell has an intermediate branching pattern. These localized differences within a single cell type may help to explain some of the widely varied arbors reported for T cells (Nicholls and Purves 1970; Van Essen 1973; Miyazaki and Nicholls 1976; Muller and McMahan 1976; Yau 1976). When two T cells within a ganglion are stained with separate markers, such as Lucifer Yellow and HRP (Macagno et al. 1981), it is clear that their secondary processes extend together into the neuropil and elaborate varicosities at similar locations (DeRiemer and Macagno 1981).

Diffusion apparently limits the spread of intracellularly injected HRP within processes of the cell. As a rule, spread of HRP into the processes within the ganglion is complete within an hour or two, but to stain the entire cell including its peripheral and intersegmental processes may require more than a day.

In this case, chains of ganglia must be kept in culture medium. After these longer incubations, the branching patterns for T, P, and N cells in adjacent ganglia (Yau 1976) are seen to be quite similar to those within the cell's own ganglion. Staining of two T cells in single preparations shows that those axons extending to adjacent ganglia elaborate branches with varicosities near those of other T cells. The

FIGURE 2

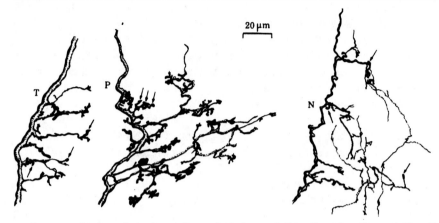

▲ Shapes of secondary branches of T, P, and N cells. The T- and N-cell processes exhibit isolated enlargements and irregularly shaped stubby fingers (↓) along their length that are sites of synapses. The P cell also has these specializations, but they are often situated in large clusters. Camera lucida drawings of cells injected with HRP and then stained with DAB and H_2O_2. (Reprinted, with permission, from Muller and McMahan 1976.)

physiological effectiveness of synapses in adjacent ganglia is well documented (Jansen et al. 1974). These synapses presumably add to the synaptic bombardment on neurons within the distant ganglia when the accessory receptive fields are stimulated, even if conduction block at branch points prevents impulses from propagating into the sensory cell's principal ganglion (see Chapter 5).

The intracellular features that allow one to distinguish chemical synapses are clearly visible in HRP-injected sensory cells stained with diaminobenzidine (DAB). Serial thin sections cut through regions of neuropil containing stained sensory cell processes reveal that all chemical synapses made by mechanosensory cells occur at varicosities on the secondary and primary processes within the neuropil (Muller and McMahan 1976). Sensory neurons also are postsynaptic to other sensory cells and unidentified neurons at varicosities. The fine structure of each sensory cell's synapses conforms to the general picture that we have drawn for chemical synapses in the leech. A presynaptic band of high electron density apposes the postsynaptic processes, which are generally paired, across a synaptic cleft. If each dense band is by convention termed a synapse, then each varicosity or cluster of fingers makes several dozen synapses upon other neurons and is postsynaptic at a few synapses. Such synaptic terminals that combine inputs and outputs are common in the mammalian retina (Dowling 1970) and are increasingly being seen elsewhere in the brains of vertebrates and invertebrates (Rakic 1975). The possibilities for integration by local circuits are intriguing but not yet tested.

In contrast to the sensory neurons, the cell bodies of most motor neurons lie contralateral to the segmental nerves into which their axons project (Chapter 5). As these efferent axons cross the ganglion on their way to the body wall musculature, they bristle with branches (Stuart 1970; Muller and McMahan 1976). Impulses are initiated in these axons as they leave the ganglion and do not invade the cell body or branches within the ganglion. The motor neurons that innervate large longitudinal (L), annulus erector (AE), and ventrolateral circular (CV) musculature are among those motor neurons that physiological studies have shown are chiefly postsynaptic to mechanosensory neurons within the ganglion (see also below). In these cells, the highly branched secondary processes that extend sometimes scores of micrometers from the axon are smooth and free of varicosities. Using Procion Yellow as a marker to locate processes of specific cells in the light microscope and correlate them with profiles seen in the same sections in the electron microscope, Purves and McMahan (1972, 1973) found that the L motor neuron is postsynaptic at presumably hundreds of synapses on spines within the neuropil. Analysis by electron microscopy, using HRP as a marker, shows that the fine branches or spines of the motor neuron processes are indeed postsynaptic to structures that resemble sensory neuron terminals. Whereas several spines of a single injected motor neuron may be postsynaptic to a single terminal, such spines will not be adjacent to each other. Some adjacent postsynaptic spines may stem from contralateral homologs, which are known to receive common inputs (Nicholls and Purves 1970; Muller and Nicholls 1974).

So far, in every case that has been examined, varicosities in the neuropil are sites of synapses. Another interneuron, the heart interneuron HN(3), which makes inhibitory synapses upon the heart excitor neuron HE(3), conforms to this picture. Typical synapses with agranular vesicles are seen at HN(3) varicosities (Fig. 3). The dense packing of synaptic vesicles and the sheetlike fins extending along the axon of this cell within the neuropil serve to illustrate some of the variety that can be found at the fine-structural level among cells of different modalities. The motor neuron inhibitory to the dorsal longitudinal muscle designated cell 1 (Appendix D), which also makes inhibitory connections within the ganglion, has a similar synaptic fine structure at swellings along and at the tips of branches (Muller 1979).

FIGURE 3

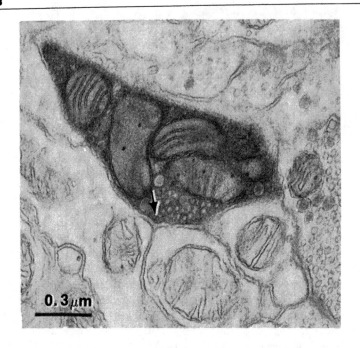

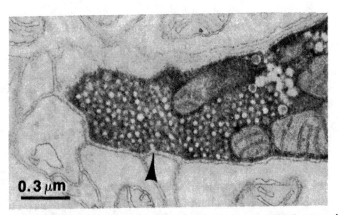

▲ Presynaptic terminals of HN cell that was injected with HRP in ganglion 3 — HN(3). (*Top*) HN(3) makes a typical synapse (↓) on unidentified cells. (*Bottom*) The synaptic vesicles are densely packed within the terminal. One vesicle is continuous with the presynaptic membrane (▲). The HN cells make inhibitory chemical synapses on other HN cells and on HE motor neurons.

Whether the transmitter contained in such terminals is the same as that at chemical synapses of sensory or S cells is unknown, but it has been shown in the leech that single neurons may be hyperpolarized at some receptors by acetylcholine (ACh) and depolarized by ACh at others; therefore the postsynaptic receptors may determine whether a particular transmitter is excitatory or inhibitory (Sargent et al. 1977). Such dual actions of one transmitter on single neurons have been well studied in the marine mollusc *Aplysia* (Ascher 1972; Kehoe 1972; Gerschenfeld 1973; Kandel 1976; Gardner and Kandel 1977) and in *Navanax* (Levitan and Tauc 1972). Although in some other systems inhibitory synaptic terminals are distinguished by containing flattened or elliptical vesicles, which have not been seen in the leech, special methods of fixation are required to retain (or create) such vesicle shapes (Tisdale and Nakajima 1976).

In leech segmental ganglia, the two large Retzius (R) cells and also five other neurons stain with Neutral Red dye (Stuart et al. 1974; Appendix D) and, on the basis of formaldehyde-induced fluorescence and direct analysis, contain 5-hydroxytryptamine (5-HT; serotonin) (see Chapter 8). The processes of these cells ramify within the neuropil and contain 50-nm granular vesicles and large and extremely dense-cored vesicles within varicosities, but no processes have been seen that make well-defined synapses.

The synapses at neuromuscular junctions in the leech are less well characterized than those at other neuromuscular junctions, but published micrographs of neuromuscular junctions (Tulsi and Coggeshall 1971; Rosenbluth 1973; Yaksta-Sauerland and Coggeshall 1973) are consistent with many of the features of chemical synapses between neurons, and the physiology of neuromuscular transmission is typically cholinergic (Kuffler 1978). In the leech, 5-HT has inhibitory effects on neuromuscular transmission and muscle contraction; terminals containing 50-nm granular vesicles, such as those in the Neutral Red-staining cells, are in close proximity with muscle and peripheral nerve and may even synapse there (Tulsi and Coggeshall 1971). Unfortunately, the known inhibitory motor neurons do not contain 5-HT or stain with Neutral Red, so the source of the granular terminals is unknown although it is probably the R cell (Mason et al. 1979).

Because postsynaptic motor neuron processes can be readily distinguished from presynaptic sensory neuron varicosities, it has recently been possible to localize synapses between particular sensory and motor neurons within the ganglion (Macagno et al. 1981) (Fig. 4). For this purpose, P cells and L motor neurons were injected with HRP and

FIGURE 4

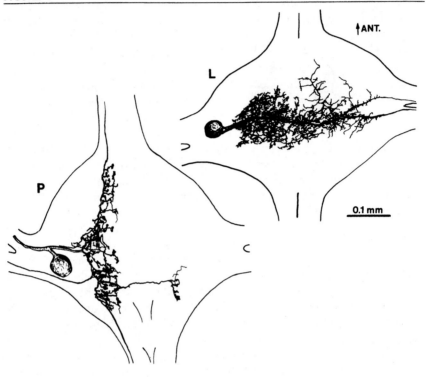

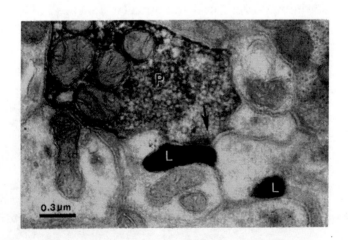

thick (8 μm) serial sections through the ganglion were taken and examined under the light microscope. The stained processes of the neurons seen in the sections were then reconstructed by computer graphics, and regions of intercellular contact were chosen for the examination of synaptic ultrastructure. The thick sections were resectioned, and the synaptic contacts were located on the thin sections under the electron microscope. The regions of the neuropil examined so far have been near the fork of the main axon of the L motor neuron, where impulses may be initiated and therefore where some of the physiologically most potent contacts might be expected to occur. Judging from the distribution of probable contacts detected by the computer and from the results of physiological studies employing conduction block at central branch points (see Chapter 5 and below), synaptic contacts are likely to be distributed within the neuropil rather than located at one region.

Electrical Synapses

To demonstrate electrical synapses within the ganglion, currents may be passed between two cells and coupling potentials monitored with intracellular electrodes. Certain morphological techniques have also been used to demonstrate direct electrical connections between cells in the leech, such as the passage of ^{3}H-labeled fucose between R cells (Rieske et al. 1975) (although these experiments were complicated by the proximity of the cell bodies) or the movement of Procion Yellow between S cells (Frank et al. 1975; Scott and Muller 1980). But the conclusive morphological demonstration of an electrical synapse, which is considered to be the gap junction, is not possible in the electron microscope without resorting to such specialized techniques as freeze fracture or by bathing tissue in lanthanum salts (see reviews by Staehelin 1974; Larsen 1977). As shown by the transmission electron microscopy of the thin sections, electrical junctions are characterized by greatly reduced intercellular space and are often tens of

◀ P cells synapse upon L cells. (*Top*) Camera lucida tracings of individual P and L cells injected with HRP, stained, and drawn from whole mounts show the arborizations of the cells within the neuropil. Synapses are made in the region of overlap in the ipsilateral central and anterior neuropil. (*Bottom*) P cell and L cell in the same ganglion were injected with HRP. Typical chemical synapse (↓) was found in section that in light microscopic reconstruction had a contact between cells (see text).

nanometers long. Because tight junctions can have a similar appearance, several techniques must be used to establish a presumed electrical junction. Coggeshall (1974) identified in thin sections gap junctions between packet and neuropil glial cells, which are electrically coupled (Kuffler and Potter 1964). Fernandez (1975) has reported that conical particles line the cytoplasmic surface of the membrane of postsynaptic processes at the synapse and has suggested that these apparent junctions between the processes are electrical junctions. With further evidence, this proposal is attractive because certain of the adjacent postsynaptic processes do belong to cells that are known to be electrically coupled, such as the L motor neurons.

To obtain better definition of electrical synapses between neurons in the leech, one can turn from the complexly interwoven neuropil to the junctions between the S cells, which we now know to be coupled through electrical synapses located in the connectives midway between ganglia, where few if any other neurons make synapses (Carbonetto and Muller 1977; Muller and Carbonetto 1979). Procion Yellow travels between S cells in adjacent ganglia (Frank et al. 1975) but, perhaps because of its low fluorescence, it does not seem to pass between other leech neurons that are electrically coupled, e.g., the R cells (K. J. Muller, unpubl.) or the T cells (Van Essen 1973). Other small, mobile fluorescent dyes such as Lucifer Yellow (Stewart 1978) or 6-carboxyfluorescein diffuse from one S cell to the next, and their high fluorescence yield reveals that they also pass specifically into cells within the ganglion that are coupled directly to the injected cell (Muller and Scott 1981; Stewart 1981). In other systems, too, passage of small dye molecules is diagnostic, but the dyes do not invariably diffuse well enough between electrically coupled cells to be detectable (Bennett 1973). Evidence that the S cells are not syncytial, but separated by a semipermeable membrane, is provided by HRP; HRP diffuses rapidly down the S-cell axon toward the adjacent ganglion, stopping abruptly about midway between ganglia. It is possible to locate adjacent S-cell axons and the contacts they make with each other because the S-cell axon is the largest in the connectives and occupies a characteristic position. The identification of contacts can be aided by injecting one cell with HRP. The two cells contact each other where the extracellular space at such contacts is narrowed to a uniform 4–6 nm, as at electrical junctions in many other invertebrates (Pappas et al. 1971; Peracchia 1973a,b). Using lanthanum salts as an extracellular tracer, there appear to be bridges of 18–20-nm spacing within this region (K.J. Muller and B.E. Thomas, unpubl.). Although S cells make typical chemical synapses within the ganglion, none has been found within the connective. S cells in the first three and last four free segmental ganglia do, however, send axons along the regionally shorter connectives into adjacent ganglia, where they form chemical synapses with unknown targets as well as electrical synapses with adjacent S cells. At a few S-cell synapses midway in the connective, there are also terminals of as yet unidentified cells that are distinguished by the presence of 50–60-nm granular vesicles. These are apparently some of the terminals in the connectives that exhibit glyoxylic acid-induced histofluorescence characteristic of 5-HT.

Physiological evidence indicates that electrical junctions in the leech may be abundant and varied. To identify particular electrical synapses within the ganglion will probably require new techniques, such as injecting coupled cells with two different substances, one of which crosses junctions and then reacts with the other to form a precipitate at the junction. We have yet to learn whether nonrectifying junctions, such as those between homologous R cells, L or AE motor neurons, or S cells, are different from the rectifying junctions between T cells and L or S cells or P cells and L motor neurons or whether nonrectifying junctions are simply antiparallel arrangements of rectifying junctions. Rectifying junctions might be expected to pass charged dye molecules better in one direction than in the other, but there is no evidence to substantiate this.

GENERAL PHYSIOLOGY OF THE SYNAPSES

Characterization of Synapses

With sensory and motor cells both present in each ganglion, conditions in the leech are particularly favorable for using electrophysiological techniques to examine synaptic contacts that underlie simple reflexes and more elaborate behavior. The electrical length constants of leech neurons seem to be unusually large (see Fig. 7 and Frank et al. 1975), allowing distant synaptic events to be measured in the cell body with a microelectrode. This same electrode can be used to excite the cell to fire impulses or otherwise activate its synapses. Records from intact (Gardner-Medwin et al. 1973; Magni and Pellegrino 1978) or semiintact (Kristan and Calabrese 1976) preparations are similar to those from isolated ganglia or chains of ganglia bathed in physiological saline, indicating that ganglia in vitro can function as they normally do even with regard to such complex behaviors as swimming (Stent et al. 1978). The underlying neuronal connections can therefore be studied outside the animal and this facilitates stable, pairwise intracellular recording. Once a synaptic connection between two cells has been demonstrated, it is of crucial importance to determine whether the connection is monosynaptic or polysynaptic, and if monosynaptic, whether it is chemically or electrically mediated.

A range of standard tests for the presence of a chemical synapse can be applied in the leech. Release of chemical transmitter requires Ca^{++} and is antagonized by Mg^{++}. In general, therefore, chemical synaptic transmission is greatly reduced or eliminated when 20 mM $MgCl_2$ replaces 20 mM NaCl in the saline bath (Fig. 5) (Nicholls and Purves 1970). This effect of Mg^{++} is counteracted by concomitantly increasing the concentration of $CaCl_2$ from 1.8 mM to 15 mM, unless there is a second intervening synapse made by an excitable interneuron. This is presumably because Ca^{++} and Mg^{++} act antagonistically in releasing transmitter but otherwise act in concert to reduce the excitability of intervening neurons in a polysynaptic pathway (Nicholls and Purves 1970). Persistent transmission in the high Mg^{++} concentrations has recently been observed at certain inhibitory chemical synapses (Nicholls and Wallace 1978b, and see below). Thus, a series of criteria

FIGURE 5

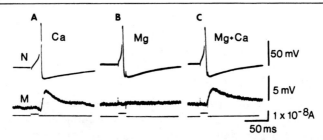

▲ Antagonistic effects of Mg^{++} and Ca^{++} on chemical synaptic transmission between the N cell and the L cell. The N cell was stimulated to give single impulses (*upper traces*) by currents injected through the microelectrode (monitored in *bottom traces*), while synaptic potentials were recorded intracellularly from the L motor neuron (M, *middle traces*). (A) The synaptic potential recorded when Ringer's fluid containing no Mg^{++} flowed past the ganglion. The Ca^{++} concentration was 20 mM to reduce the incidence of spontaneous synaptic potentials on the motor neuron. (B) Taken 8 min after changing solutions, the synaptic potential was abolished in Ringer's fluid containing 18 mM Mg^{++} and 1.8 mM Ca^{++}. Almost complete recovery occurred when the Ca^{++} concentration was increased to 15 mM, even though the Mg^{++} concentration remained 18 mM. (C) Taken 3 min after changing fluids, this is the result expected for a chemical synapse. (Modified from Nicholls and Purves 1970.)

must be brought to bear on any synapse under analysis. Specific blockers of known transmitters have been used to identify chemical synapses and to distinguish between transmitter candidates, but such agonists and antagonists tend to act more slowly than Mg^{++}, presumably because of diffusion barriers into the neuropil (Sargent et al. 1977).

Electrical synapses can be identified chiefly by two criteria. First, postsynaptic potentials (PSPs) follow the presynaptic impulse with no measurable delay. Second, because electrical synapses are by definition connections across which current can pass in at least one direction, they can be confirmed with the direct passage of current between pre- and postsynaptic neurons. Some PSPs have electrical and chemical components; these may be separated by cooling the preparation, which delays the chemical component (Fig. 6) (Nicholls and Purves 1972).

A "chemical" synaptic potential can with confidence be classed as monosynaptic only by applying several criteria, including the abovementioned persistence in solutions with high Ca^{++} and Mg^{++}. In addition, a short, constant synaptic delay has been a useful, but not absolutely reliable, indicator for monosynaptic connections within the ganglion, where the time for conduction of impulses to the synapse is

FIGURE 6

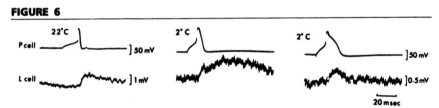

▲ Effect of cooling on the combined chemical and electrical synapse between a P cell (*upper trace*) and an L cell (*lower trace*). At 22°C, the presynaptic action potential evokes a synaptic potential that begins with no appreciable delay. Part of this potential is depressed by high Mg^{++} and augmented by high Ca^{++} (see Nicholls and Purves 1970). On cooling to 2°C, a notch appears between the two components, presumably because of an increase in latency at the chemical synapse. Subsequently, with stimulation at 1/sec for a few seconds, the delayed chemical component disappeared, leaving only the electrically mediated coupling potential. The presynaptic action potential increased in duration with repetitive stimulation in the cold. All these changes were reversed by rewarming the preparation to room temperature. In this experiment, the bathing fluid contained 15 mM Ca^{++}. (Reprinted, with permission, from Nicholls and Purves 1972.)

minimal. A chemically mediated synaptic potential is almost certainly monosynaptic if it is enhanced after injection of tetraethylammonium ions (TEA$^+$) into the presynaptic neuron. In certain neurons, the action potential lengthens gradually as the outward K$^+$ currents become blocked by TEA$^+$ (Kleinhaus and Prichard 1975, 1977). Consequently, impulses in such injected cells increase the release of transmitter in a graded fashion, with a correspondingly smooth increase in synaptic potential (Kehoe 1972; but cf. Deschenes and Bennett 1974). If, instead, an interneuron that fires nerve impulses intervenes, the synaptic potential in the observed cell grows as a staircase. Each step is produced by one interneuronal impulse in a burst generated by the growing synaptic potential at the first synapse. Two less easily applied tests of monosynaptic contacts also depend upon changing the size of the synaptic potential in a graded rather than discrete manner, either by shortening the presynaptic impulse with an appropriately timed, huge pulse of negative current or by potentiating transmitter release with a negative-going prepulse through the stimulating microelectrode (Nicholls and Purves 1970; Muller and Nicholls 1974). It is satisfying that the techniques have so far produced consistent results. None, however, has been able to eliminate entirely the possibility of participation of an unknown interneuron that intervenes between electrically coupled cells (see below and Muller and Scott 1981). A related situation that is amenable to analysis occurs when a postsynaptic neuron is electrically coupled to another identified neuron that is also postsynaptic to the common presynaptic input. Bowling et al. (1978) have shown that one postsynaptic cell can be eliminated without altering the synaptic input to the other simply by selectively killing one cell with intracellular injection of protease (Parnas and Bowling 1977). This indicates that the synaptic input was direct and not via the eliminated cell, a fact that would otherwise be difficult to prove.

There is some variability in the size of the synaptic interactions that can be recorded with microelectrodes in particular pairs of neurons, but outside a range of a few millivolts, artefacts of the recording situation, such as increased leakage conductance of the cell membrane, account for the observed variations. To improve the quality of recordings, chemical synaptic potentials may be enhanced twofold or more by bathing in 8 mM CaCl$_2$ saline solution rather than in the normal solution containing 1.8 mM Ca^{++} (Nicholls and Purves 1970). The raised Ca^{++} not only enhances transmitter release, but additionally helps to seal the cell membrane around the microelectrode and, by reducing impulse excitability, lowers the level of background synaptic noise. Therefore, despite some variation in their strengths, the specific connections between sensory and motor cells are present consistently, and one can profitably examine the conditions under which the synaptic potentials change with use and disuse, either during trains of presynaptic impulses or after lesions to the nervous system.

Synapses between Particular Neurons

In the leech, the synaptic connections between particular pairs of neurons have the high degree of regularity that characterizes the modalities, receptive and motor fields, and positions of individual cells within the ganglion. The description of chemical and electrical synaptic interactions that follows is not meant as a catalog of the synaptic connections in the medicinal leech, but rather as representative of types of synapses that have been particularly well studied and that will be valuable later in the consideration of regeneration. Many of the same connections have been found in other species of leech (see, e.g., Kramer 1981).

One of the first synapses to be described in the leech was the nonrectifying electrical synapse (across which current can pass in both directions) between R cells (Hagiwara and Morita 1962; Eckert 1963). This finding set the stage for the discovery of electrical coupling between certain symmetrically placed homologous neurons within the ganglion (Baylor and Nicholls 1969c), particularly between motor neurons that act in concert and so tend to fire impulses in synchrony (Stuart 1970). Electrical synapses exist between glial cells and presumably are similar to those between neurons (Kuffler and Potter 1964). Attempts to uncouple electrical synapses in the leech with propionate replacing Cl^- or with Ca^{++} chelaters have not yet succeeded (Payton and Loewenstein 1968), nor have experiments been performed under conditions that are known to alter intracellular pH.

Another well-studied electrical junction is that between S cells in adjacent ganglia. Even though the S cells are separated at their cell bodies by nearly 1 cm in adult animals, current passes well from one cell to the next. A signal (hyperpolarizing or depolarizing) generated in one S cell is still at least one-tenth its original size when it reaches the neighboring S cell soma (Fig. 7) (Frank et al. 1975; Muller and Carbonetto 1979). The S cell has been reported to be excited by rectifying electrical synapses that allow current to enter it from T cells (Gardner-Medwin et al. 1973), although the connection is mediated by "coupling interneurons" (Appendix D) that make nonrectifying junctions with the S cell and rectifying junctions with T cells (Muller and Scott 1981). These properties have made the S cell a most useful neuron for studying the morphology and regeneration of electrical junctions, as described in Chapter 10.

FIGURE 7

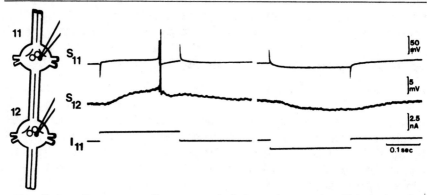

▲ S cells in adjacent ganglia are coupled through a nonrectifying electrical synapse. The S cell in ganglion 11 (*top traces*) was depolarized and hyperpolarized by currents injected through the microelectrode (monitored in *bottom traces*). Current spread about 5 mm into the S cell in the adjacent ganglion (*middle trace*). Current injected into the S cell in ganglion 12 flowed equally well into the S cell in ganglion 11 (not shown). A schematic diagram of the recording configuration is shown at the left. S cells synapse with each other midway between ganglia in the medial connective.

The symmetry of electrical coupling between two cells, i.e., whether potentials spread equally from one cell or the other, is known to depend not only on resistance of the electrical synapse between the cells, but also on their individual resistances. The rostral and lateral penile evertor (PE) motor neurons (Zipser 1979a; Chapter 5), for example, are coupled by a nonrectifying electrical synapse (Zipser 1979b). Zipser (1976b) has found that because the rostral neuron's membrane resistance greatly increases on depolarization, however, depolarizing potentials in the lateral neuron can spread effectively into the rostral motor neuron, but not vice versa. This connection is distinct from a rectifying junction in its mechanism, though not necessarily in its effect.

The largest literature on synaptic physiology of the leech deals with synapses made by sensory neurons, particularly onto motor neurons. The cell bodies of sensory neurons are electrically excitable, so that direct electrical stimulation with an intracellular microelectrode mimics natural stimulation and effectively activates synapses. In contrast, motor neuron cell bodies, and presumably dendrites, do not fire action potentials and thus serve as excellent regions over which signals may be integrated. The efferent impulses produced in the motor neuron by a sufficiently excitatory stimulus are generated at a distance from the cell body. T and P mechanosensory neurons make rectifying electrical synapses with the L motor neuron that activates the longitudinal musculature, thus shortening the segment (Nicholls and Purves 1970). In this case, excitatory currents can pass from the sensory cells into the motor neuron, but only negative currents can pass in the opposite direction (so that the animal does not feel that it is being touched or pressed every time the L motor neuron fires an impulse). There is no evidence that at these synapses the transmission of hyperpolarization has functional significance, as it does at another synapse, namely, that between two neurons in the swimming oscillator, where an electrical synapse is selectively inhibitory (Friesen et al. 1976, 1978; Stent et al. 1978; see Chapter 7). The most unusual electrical synapse yet described in the leech is a double-rectifying synapse that allows positive but not negative currents to pass in either direction between T cells (Baylor and Nicholls 1969c). A variant of Baylor and Nicholls' original explanation for this phenomenon still seems plausible — that channels between cells at electrical junctions increase their resistance with intracellular hyperpolarization and therefore exhibit the same rectifying characteristics as certain nonjunctional channels. They also considered that an intervening interneuron, and therefore two serial junctions, could be involved. This does, in fact, explain the connection between T cells on opposite sides of the ganglion, which are linked via rectifying junctions with the coupling interneurons; but selective destruction of the coupling interneurons does not uncouple ipsilateral T cells (Muller and Scott 1981).

By the criteria outlined above for characterizing chemical synaptic potentials, P and N cells produce monosynaptic excitatory postsynaptic potentials (EPSPs) in the L motor neurons (Nicholls and Purves 1970). These same sensory cells produce similar but somewhat less powerful monosynaptic EPSPs in the AE motor neurons (Muller and Nicholls 1974). These connections help to account for the reflex shortening and erection of annuli when the leech is squeezed or pinched.

Some of the sensory–motor connections that have been examined within the ganglion have also been shown to exist between sensory cells in one ganglion and motor cells in the next. For example, the P and N cells project axons into adjacent ganglia to synapse directly upon neighboring L motor neurons. For such connections between cells in separate ganglia, the only potent physiological test for monosynaptic connections that can readily be applied is the persistence of synaptic potentials in saline with raised Mg^{++} and Ca^{++} concentrations (Jansen et al. 1974).

A series of inhibitory chemical synapses has been discovered to control the neurogenic blood circulation in the leech (Thompson and Stent 1976b). HN cells rhythmically inhibit certain ipsilateral HE motor neurons, most of which are in distant ganglia, and inhibit contralateral homologous HN cells, a case of presynaptic inhibition. Only in the third segmental ganglion do both of these interactions occur entirely within the ganglion. Nicholls and Wallace (1978a,b) have studied transmitter release and the interactions at this synapse, and the following account derives from their study.

The inhibitory postsynaptic potentials (IPSPs) in the HE motor neuron reverse polarity when the membrane potential is shifted more negative than −75 mV, and they are produced by an increase in Cl^- conductance (Fig. 8). Under normal recording conditions, there is apparently some steady release of inhibitory transmitter from the HN cell, and this release can be arrested by hyperpolarizing the cell.

Several lines of evidence indicate that transmitter is released in quantal units at this synapse, although it is not possible under physiological conditions to record unambiguous miniature synaptic potentials. As shown in Fig. 3, HN cell terminals contain 50-nm synaptic vesicles densely clustered at apparently typical chemical synapses. Nicholls and Wallace reduced or completely eliminated other synaptic inputs to the HE cell by replacing Na^+ in the bath, mole-for-mole, either with 20 mM Mg^{++} or isotonic sucrose. In the latter case, when the unitary events that look like miniature synaptic potentials in the HE motor neuron were reversed in sign and enhanced by injecting Cl^- intracellularly through the recording electrode, a depolarization of the HN cell increased the frequency of the discrete events. Because the presynaptic neurons cannot fire impulses under such conditions, it was not possible to demonstrate conclusively that these quantal events sum to make up the full synaptic potential. However, under more nearly physiological conditions, quantal unit size was determined in two independent ways. First, while producing a series of IPSPs in the HE motor neuron with impulses in the HN interneuron, the mean IPSP amplitude was determined and the number of synaptic failures was estimated (Figs. 9 and 10). Assuming quantal release according to Poisson statistics, the unit size was then calculated as: (mean IPSP amplitude)/ln (number of trials/number of failures), and was found to be about one-quarter of a millivolt. Second, because a sustained depolarization of the presynaptic HN cell causes a steady release of transmitter, an analysis of postsynaptic noise could reveal the amplitude and time course of quantal events. Nicholls and Wallace used noise-analysis techniques like those that have been applied to quantal analysis in the visual system (Dodge et al. 1968) and at motor end-

FIGURE 8

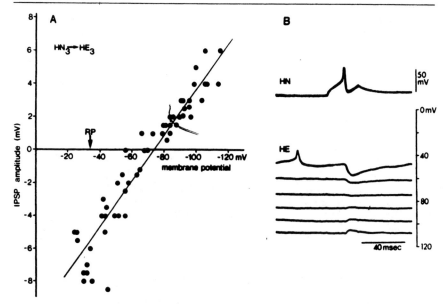

▲ Effect of membrane potential of postsynaptic HE cell on IPSP amplitude. (A) The HN cell was stimulated once every 2 sec, and the amplitude of the synaptic potential evoked in the HE cell was varied by passing steady currents through the recording microelectrode. (B) Intracellular recordings from the HN cell and HE cell from a similar experiment. Preparations were bathed in Ringer's saline containing 7.5 mM Ca^{++}. (Reprinted, with permission, from Nicholls and Wallace 1978a.)

plates (Katz and Miledi 1970, 1972); they assumed that the recording electrode measures events of uniform amplitude and rapid rise-time that sum linearly and occur at random times to produce an average depolarization V and an accompanying variance of the signal $\bar{E}^2$. For Poisson processes, the unit amplitude is $2\bar{E}^2/V$, which was calculated to be between one-fifth and one-quarter of a millivolt. This agreement, by the two distinct methods, reinforces the assumption that release is quantal and at low levels obeys Poisson statistics. The one-quarter-millivolt unit size was not increased when greater numbers of quanta were released but was 10–20% larger in the postsynaptic (contralateral) HN cell than in the HE neuron.

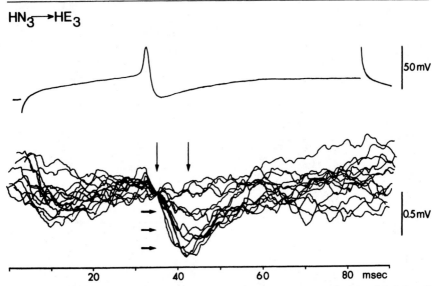

▲ Fluctuations in amplitude of 14 consecutive IPSPs recorded in the HE cell and evoked by stimulation of the HN cell. The traces were aligned horizontally by superimposing the peaks of the HN action potentials and vertically by superimposing the responses in the HE cell at a point 2 msec after the peak of the action potential (the point of minimum latency). The amplitude of each synaptic potential was measured as the potential difference between the point of minimum latency and the point corresponding to the peak of the averaged response (↓). (→) Integral multiples of the amplitude of the quantal unit determined from the number of failures by Poisson statistics (see text). The amplitude distribution of all the responses from this pair of cells is shown in Fig. 10A. In this experiment, the apparent electrical coupling between HN and HE cells was an artefact due to direct interaction between the electrodes. The Ringer's fluid contained 20 mM $MgCl_2$. (Reprinted, with permission, from Nicholls and Wallace 1978b.)

Short-term Changes in Synaptic Transmission

The pattern of impulses ordinarily used in the preceding studies to characterize particular synapses is in a sense artificial, usually consisting of isolated impulses separated from each other by several seconds. In normally functioning cells, impulses frequently come in bursts. It has been known for some time that at peripheral chemical synapses, impulses within a train produce varied effects, depending upon the

FIGURE 10

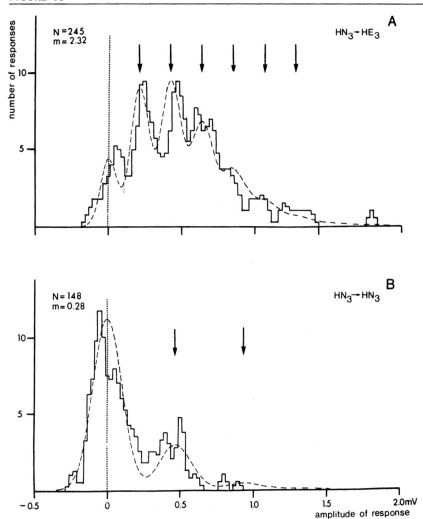

▲ Amplitude histogram of IPSPs evoked by stimulation of an HN cell. The Ringer's fluid contained 20 mM $MgCl_2$. (———) The observed T responses (with the sign reversed and three-point averaged). (---------) The distributions predicted by Poisson statistics, estimating the quantal content, m, from the number of failures and arbitrarily assigning a standard deviation to the unit. (↓) Integral multiples of the unit amplitude. (A) Distribution of responses in an HE cell: Unit amplitude = 0.213 mV, $\sigma_{background}$ = 55 μV, σ_{unit} = 30 μV. (B) Distribution of responses in a contralateral HN cell: Unit amplitude = 0.47 mV, $\sigma_{background}$ = 100 μV, σ_{unit} = 40 μV. (Reprinted, with permission, from Nicholls and Wallace 1978b.)

past history of stimulation, the firing frequency, and the extracellular Ca^{++} concentration (del Castillo and Katz 1954; Dudel and Kuffler 1961; Hubbard 1974). Typically, successive synaptic potentials in a train grow in amplitude (facilitate) until a peak amplitude of individual synaptic potentials is reached. Successive PSPs may then diminish, in some cases below the initial level (depression). At neuromuscular junctions in general, facilitation and depression come about because more and then fewer quanta are released from the presynaptic terminal; these events occur more quickly with high firing rates or high Ca^{++} concentration. Another effect, posttetanic potentiation, lasts for seconds or even minutes after a train of impulses; during this time a single impulse can produce an increased synaptic potential. Nicholls and Purves (1972) demonstrated similar facilitation, posttetanic potentiation, and depression at P- and N-cell chemical synapses upon the L motor neuron. In contrast, at the electrical synapse of the T and P cells with the same L motor neuron, successive synaptic potentials within a train were seen to be of uniform amplitude. Depression was not observed, even at high presynaptic firing rates. Electrical transmission is therefore more constant than chemical transmission, not only for individual PSPs, but also with regard to PSPs in a train. This indefatigability both at high frequencies and at low temperatures provides electrical synapses with a good deal of reliability.

Another aspect of facilitation and depression, which was first seen at a peripheral neuromuscular junction in Crustacea (Atwood and Bittner 1971; Frank 1973), is that a motor neuron that synapses on two or more muscle fibers can produce different amounts of facilitation and depression at different synapses. Under some conditions during a train of impulses, one terminal of the neuron can release increasingly more quanta of transmitter while another branch of the same cell might release successively fewer. In these cases, it is characteristic of the synapses at which there is an initially small synaptic potential to exhibit facilitation, whereas at branches having more powerful synapses there is an earlier onset of depression. It is also typical that all the inputs to a given muscle fiber exhibit a similar ability to produce facilitation. Muller and Nicholls (1974) compared facilitation and depression at synapses of single P and N sensory cells onto both the L and AE motor neurons and found just such differential effects between neurons (Fig. 11). Synapses of the sensory neurons upon the L motor neuron are far more potent than synapses of the same neurons upon the AE motor neuron. As one might therefore predict, EPSPs in the AE motor neuron showed more facilitation with less depression than in the L motor neuron. This effect may well account for a difference in the actions of the two motor neurons that can be ob-

FIGURE 11

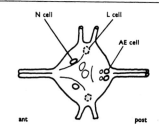

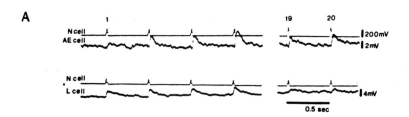

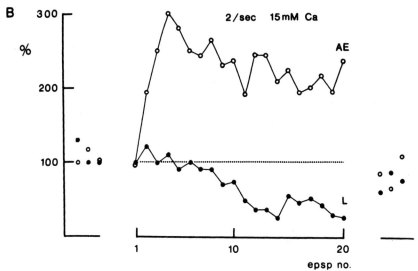

▲ Trains of impulses in N cells in fluid containing 15 instead of 1.8 mM Ca^{++}. (A) Records from AE and L cells at the beginning and end of trains at 2/sec. (B) Whereas the synaptic potentials in the AE cell (o) were facilitated, those in the L cell (•) were depressed. Intervals between tests were 30 sec before the train and 10 sec after the train. (---------) The average value of synaptic potentials before the train. (Reprinted, with permission, from Muller and Nicholls 1974.)

served behaviorally — pinching the animal can cause it to shorten transiently with a slower, longer-lasting erection of annuli.

Facilitation and depression are not properties of all chemical synapses in the leech. Specifically, the inhibitory synapses made by the HN cell do not exhibit classical facilitation or depression (Thompson and Stent 1976b; Nicholls and Wallace 1978a,b), although the IPSPs produced by the HN cell are modifiable. A presynaptic depolarization above resting potential, such as produces a train of impulses, causes

FIGURE 12

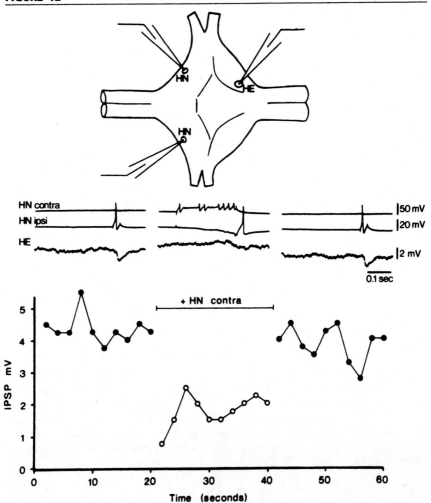

synaptic potentials to increase with a time course of about a half second, much like facilitation (Thompson and Stent 1976c; Nicholls and Wallace 1978a,b). Conversely, a presynaptic hyperpolarization in one HN cell, whether produced with extrinsic current or IPSPs, reduces the amount of transmitter released. Therefore, when one HN cell is caused to inhibit another HN cell, it is a case of presynaptic inhibition (Fig. 12).

A different activity-dependent reduction of transmitter release has been reported by Yau (1975). P sensory neurons extend afferent axons through adjacent ganglia to the periphery much as T cells do. Impulses that enter the ganglion along the P cell's fine axon in the posterior connective normally propagate through the ganglion and out along the cell's other axons. During an intense burst of impulses and for some time thereafter, impulses may fail to penetrate the ganglion from the connective, stopping where the small connective axon branches from the larger root axon. The cause of branch-point failure is simply that action potentials generated anywhere in the cell propagate through the major axons, activating an electrogenic Na^+ pump, or K^+ conductance increase, or both in the cell membrane, hyperpolarizing it, and making it more difficult to fire impulses (Baylor and Nicholls 1969a,b; Jansen and Nicholls 1973; Van Essen 1973). The smaller currents generated in the connective axon, unable to depolarize the larger axons, become the first to fail. When the P cell's impulse fails at the posterior branch-point, the synaptic potential in the L cell disappears. This suggests that the strongest synapses onto the L cell do not arise from the P cell's posterior process. On the other hand, when an impulse travelling into the ganglion from the anterior connective fails at an anterior branch-point, 20–30% of the normal synaptic potential remains (E. Macagno and K. J. Muller, unpubl.). These findings are consistent with those of Macagno et al. (1981) (Fig. 5) that

◀ Presynaptic inhibition at the inhibitory synapse between HN and HE cells. Action potentials were initiated in an HN cell once every 2 sec and the amplitude of the IPSP in the ipsilateral HE motor neuron measured. After 10 control stimuli, each impulse in the HN cell was preceded by a burst of IPSPs evoked by stimulating the contralateral HN cell. As a result of this synaptic input, the action potential in the ipsilateral HN cell arose from a more hyperpolarized membrane potential and therefore gave rise to a smaller IPSP in the HE cell. When the contralateral HN cell was not stimulated, the IPSPs in the HE cell returned to control values (15 mM Mg^{++}, 1.8 mM Ca^{++} Ringer's fluid). (Reprinted, with permission, from Nicholls and Wallace 1978a).

central P-cell processes make numerous synapses on the fine branches and spines near the main axon of the L cell, but that anterior branches of the P cell may also make synapses on the L cell. A similar reduction in amplitude of synaptic potential as a result of branch-point failure has now been seen in the rectifying electrical connection of the T cell with the S interneuron (Fig. 13) (Muller and Scott 1981). The synaptic potential in the S cell, mediated by the coupling interneurons, is reduced but not eliminated by branch-point failure in the T cell. This indicates that the connections between the T cells and coupling interneurons are spatially distributed, in agreement with the anatomy of the HRP-injected cells.

In summary, the electrical and chemical synaptic interactions, the morphology of synapses, and their short-term modifiability all have

FIGURE 13

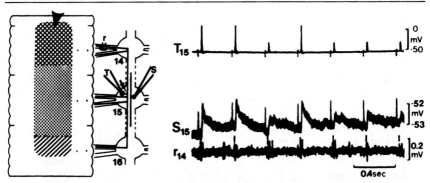

▲ Conduction block produced by natural stimulation of the minor field of the T cell reduces the amplitude of the synaptic potential in the S cell. Schematic diagram shows the approximate location of the stimulatory stylus (▼) in the anterior minor field (▓) of the lateral T cell. Simultaneous recordings were made intracellularly from the T and S cells (*top* and *middle traces*, respectively) and extracellularly with a modified hook electrode applied to the anterior root of the adjacent ganglion (*r*, and *bottom traces*). Arrow in ganglion 15 indicates site of conduction block. The preparation was bathed in Ringer's fluid containing 20 mM $MgCl_2$. Conduction block that occurs during a train of impulses at 3 Hz reduces the amplitude of the impulse recorded in the T-cell soma and the synaptic potential in the S cell. The recording from the anterior root (r) indicates that only the T cell penetrated by the microelectrode and no T cell in ganglion 14 is stimulated. A stimulus artefact precedes each intracellularly recorded impulse by approximately 40 msec. (Reprinted, with permission, from Muller and Scott 1981.)

counterparts in mammalian central nervous systems, where function and structure cannot so readily be traced to individual cells. Although the synapses described in this chapter provide a basis for simpler reflexes in the leech, they are also the elements of circuit diagrams that provide a neuronal basis for coordinated movements, as discussed in Chapter 7. How the synapses form during development and during regeneration is under intensive study (see Chapters 9 and 10).

REFERENCES

Armett-Kibel, C., I.A. Meinertzhagen, and J.E. Dowling 1977. Cellular and synaptic organization in the lamina of the dragon-fly, *Sympetrum rubicundulum*. *Proc. R. Soc. Lond. B* **196**:385–413.

Ascher, P. 1972. Inhibitory and excitatory effects of dopamine on *Aplysia* neurones. *J. Physiol.* **225**:173–209.

Atwood, H.L. and G.D. Bittner. 1971. Matching of excitatory and inhibitory inputs to crustacean muscle fibers. *J. Neurophysiol.* **34**:157–170.

Baylor, D.A. and J.G. Nicholls. 1969a. Long-lasting hyperpolarization after activity of neurons in leech central nervous system. *Science* **162**:279–281.

———. 1969b. After-effects of nerve impulses on signalling in the central nervous system of the leech. *J. Physiol.* **203**:571–589.

———. 1969c. Chemical and electrical synaptic connexions between cutaneous mechanoreceptor neurones in the central nervous system of the leech. *J. Physiol.* **203**:591–609.

Bennett, M.V.L. 1973. Function of electrotonic junctions in embryonic and adult tissues. *Fed. Proc.* **32**:65–75.

Bowling, D., J. Nicholls, and I. Parnas. 1978. Destruction of a single cell in the C.N.S. of the leech as a means of analyzing its connexions and functional role. *J. Physiol.* **282**:169–180.

Carbonetto, S. and K.J. Muller. 1977. A regenerating neurone in the leech can form an electrical synapse on its severed axon segment. *Nature* **267**:450–452.

Coggeshall, R.E. 1974. Gap junctions between identified glial cells in the leech. *J. Neurobiol.* **5**:463–467.

Coggeshall, R.E. and D.W. Fawcett. 1964. The fine structure of the central nervous system of the leech, *Hirudo medicinalis*. *J. Neurophysiol.* **27**:229–289.

Couteaux, R. and M. Pecot-Dechavassine. 1970. Vesicules synaptiques et poches au niveau des 'zones actives' de la jonction neuromusculaire. *C. R. Acad. Sci. D* **271**:2346–2349.

del Castillo, J. and B. Katz. 1954. Statistical factors involved in neuromuscular facilitation and depression. *J. Physiol.* **124**:574–585.

DeRiemer, S.A. and E.R. Macagno. 1981. Light microscopic analysis of contacts between pairs of identified leech neurons with combined use of horseradish peroxidase and Lucifer Yellow. *J. Neurosci.* **1**:650–657.

Deschènes, M. and M.V.L. Bennett. 1974. A qualification to the use of TEA as a tracer for monosynaptic pathways. *Brain Res.* **77**:169–172.

Dodge, F.A., B.W. Knight, and J. Toyoda. 1968. Voltage noise in *Limulus* visual cells. *Science* **160**:88–90.

Dowling, J.E. 1970. Organization of vertebrate retinas. *Invest. Ophthalmol.* **9**:655–680.

Dreyer, F., K. Peper, K. Akert, C. Sandri, and H. Moor. 1973. Ultrastructure of the "active zone" in the frog neuromuscular junction. *Brain Res.* **62**:373–380.

Dudel, J. and S.W. Kuffler. 1961. Presynaptic inhibition at the crayfish neuromuscular junction. *J. Physiol.* **155**:543–562.

Eckert, R. 1963. Electrical interaction of paired ganglion cells in the leech. *J. Gen. Physiol.* **46**:573–587.

Fernández, J.H. 1975. Structure and arrangement of synapses in the C.N.S. of the leech under normal and experimental conditions. *Neurosci. Abstr.* **1**:1023.

Frank, E. 1973. Matching of facilitation at the neuromuscular junction of the lobster: A possible case for influence of muscle on nerve. *J. Physiol.* **233**:635–658.

Frank, E., J.K.S. Jansen, and E. Rinvik. 1975. A multisomatic axon in the central nervous system of the leech. *J. Comp. Neurol.* **159**:1–13.

Friesen, W.O., M. Poon, and G.S. Stent. 1976. An oscillatory neuronal circuit generating a locomotory rhythm. *Proc. Natl. Acad. Sci.* **73**:3734–3738.

_____. 1978. Neuronal control of swimming in the medicinal leech. IV. Identification of a network of oscillatory interneurones. *J. Exp. Biol.* **75**:25–43.

Gardner, D. and E.R. Kandel. 1977. Physiological and kinetic properties of cholinergic receptors activated by multiaction interneurons in buccal ganglia of *Aplysia*. *J. Neurophysiol.* **40**:333–348.

Gardner-Medwin, A.R., J.K.S. Jansen, and T. Taxt. 1973. The "giant" axon of the leech. *Acta Physiol. Scand.* **87**:30A–31A.

Geffen, L.B. and B.G. Livett. 1971. Synaptic vesicles in sympathetic neurons. *Physiol. Rev.* **51**:98–157.

Gerschenfeld, H.M. 1973. Chemical transmission in invertebrate central nervous systems and neuromuscular junctions. *Physiol. Rev.* **53**:1–119.

Hagiwara, S. and H. Morita. 1962. Electrotonic transmission between two nerve cells in the leech ganglion. *J. Neurophysiol.* **25**:721–731.

Heuser, J.E., T.S. Reese, and D.M.D. Landis. 1974. Functional changes in frog neuromuscular junctions studied with freeze-fracture. *J. Neurocytol.* **3**:109–131.

Hubbard, J.I. 1974. Neuromuscular transmission — Presynaptic factors. In *The peripheral nervous system* (ed. J. I. Hubbard), pp. 151–180. Plenum Press, New York.

Jansen, J.K.S. and J.G. Nicholls. 1973. Conductance changes, an electrogenic pump and the hyperpolarization of leech neurones following impulses. *J. Physiol.* **229**:635–665.

Jansen, J.K.S., K.J. Muller, and J.G. Nicholls. 1974. Persistent modification of synaptic interactions between sensory and motor nerve cells following discrete lesions in the central nervous system of the leech. *J. Physiol.* **242**:289–305.

Kandel, E.R. 1976. *Cellular basis of behavior: An introduction to behavioral neurobiology.* W.H. Freeman, San Francisco.

Katz, B. and R. Miledi. 1970. Membrane noise produced by acetylcholine. *Nature* **226**:962–963.

————. 1972. The statistical nature of the acetylcholine potential and its molecular components. *J. Physiol.* **224**:665–699.

Kehoe, J. 1972. The physiological role of three acetylcholine receptors in synaptic transmission in *Aplysia. J. Physiol.* **225**:147–172.

Kleinhaus, A.L. and J.W. Prichard. 1975. Calcium-dependent action potentials produced in leech Retzius cells by tetraethylammonium chloride. *J. Physiol.* **246**:351–361.

————. Close relation between TEA responses and Ca-dependent membrane phenomena of four identified leech neurones. *J. Physiol.* **270**:181–194.

Kramer, A.P. 1981. The nervous system of the glossiphoniid leech *Haementeria ghilianii.* II. Synaptic pathways controlling body wall shortening. *J. Comp. Physiol.* (in press).

Kristan, W.B., Jr. and R.L. Calabrese. 1976. Rhythmic swimming activity in neurones of the isolated nerve cord of the leech. *J. Exp. Biol.* **65**:643–668.

Kuffler, D.P. 1978. Neuromuscular transmission in longitudinal muscle of the leech, *Hirudo medicinalis. J. Comp. Physiol.* **124**:333–338.

Kuffler, S.W. and D.D. Potter. 1964. Glia in the leech central nervous system: Physiological properties and neurone-glia relationship. *J. Neurophysiol.* **27**:290–320.

Larsen, W.J. 1977. Structural diversity of gap junctions. *Tissue Cell* **9**:373–394.

Levitan, H. and L. Tauc. 1972. Acetylcholine receptors: Topographic distribution and pharmacological properties of two receptor types of a single molluscan neuron. *J. Physiol.* **222**:537–558.

Macagno, E.R., K.J. Muller, W. Kristan, S. DeRiemer, R. Stewart, and B. Granzow. 1981. Mapping of neuronal contacts with intracellular injection of horseradish peroxidase and Lucifer Yellow in combination. *Brain Res.* **217**:143–149.

Magni, F. and M. Pellegrino. 1978. Patterns of activity and the effects of activation of the fast conducting system on the behavior of unrestrained leeches. *J. Exp. Biol.* **76**:123–135.

Mason, A., A.J. Sunderland, and L.D. Leake. 1979. Effects of leech Retzius cells on body wall muscles. *Comp. Biochem. Physiol.* **63C**:359–361.

McMahan, U.J. and S.W. Kuffler. 1971. Visual identification of synaptic boutons on living ganglion cells and of varicosities in postganglionic axons in the heart of the frog. *Proc. R. Soc. Lond. B* **177**:485–508.

Miyazaki, S. and J.G. Nicholls. 1976. The properties and connexions of nerve cells in leech ganglia maintained in culture. *Proc. R. Soc. Lond. B* **194**:295–311.

Muller, K.J. 1979. Synapses between neurones in the central nervous system of the leech. *Biol. Rev.* **54**:99–134.

Muller, K.J. and S.T. Carbonetto. 1979. The morphological and physiological properties of a regenerating synapse in the C.N.S. of the leech. *J. Comp. Neurol.* **185**:485–516.

Muller, K.J. and U.J. McMahan. 1975. The arrangement and structure of synapses formed by specific sensory and motor neurons in segmental ganglia of the leech. *Anat. Rec.* **181**:432.

———. 1976. The shapes of sensory and motor neurones and the distribution of their synapses in ganglia of the leech: A study using intracellular injection of horseradish peroxidase. *Proc. R. Soc. Lond.* B **194**:481–499.

Muller, K.J. and J.G. Nicholls. 1974. Different properties of synapses between a single sensory neurone and two different motor cells in the leech C.N.S. *J. Physiol.* **238**:357–369.

Muller, K.J. and S.A. Scott. 1981. Transmission at a "direct" electrical connexion mediated by an interneurone in the leech. *J. Physiol.* **311**:565–583.

Nicholls, J.G. and D. Purves. 1970. Monosynaptic chemical and electrical connexions between sensory and motor cells in the central nervous system of the leech. *J. Physiol.* **209**:647–667.

———. 1972. A comparison of chemical and electrical synaptic transmission between single sensory cells and a motorneurone in the central nervous system of the leech. *J. Physiol.* **225**:637–656.

Nicholls, J.G. and B.G. Wallace. 1978a. Modulation of transmission at an inhibitory synapse in the C.N.S. of the leech. *J. Physiol.* **281**:157–170.

———. 1978b. Quantal analysis of transmitter release at an inhibitory synapse in the C.N.S. of the leech. *J. Physiol.* **281**:171–185.

Ort, C.A., W.B. Kristan, Jr., and G.S. Stent. 1974. Neuronal control of swimming in the medicinal leech. II. Identification and connections of motor neurons. *J. Comp. Physiol.* **94**:121–154.

Pappas, G.D., Y. Asada, and M.V.L. Bennett. 1971. Morphological correlates of increased coupling resistance at an electronic synapse. *J. Cell Biol.* **49**:173–188.

Parnas, I. and D. Bowling. 1977. Killing of single neurons by intracellular injection of proteolytic enzymes. *Nature* **370**:626–628.

Payton, B.W. and W.R. Loewenstein. 1968. Stability of electrical coupling in leech giant nerve cells: Divalent cations, propionate ions, tonicity and pH. *Biochim. Biophys. Acta* **150**:156–158.

Peracchia, C. 1973a. Low resistance junction in crayfish. I. Two arrays of globules in junctional membranes. *J. Cell Biol.* **57**:54–65.

———. 1973b. Low resistance junctions in crayfish. II. Structural details and further evidence for intercellular channels by freeze-fracture and negative-staining. *J. Cell Biol.* **57**:66–76.

Purves, D. and U.J. McMahan. 1972. The distribution of synapses on a physiologically identified motor neuron in the central nervous system of the leech. *J. Cell Biol.* **55**:205–220.

———. 1973. Procion Yellow as a marker for electron microscopic examination of functionally identified nerve cells. In *Intracellular staining in neurobiology* (ed. S.B. Kater and C. Nicholson), pp. 72–81. Springer-Verlag, New York.

Rakic, P. 1975. Local circuit neurones. *Neurosci. Res. Prog. Bull.* **13**:291–440.

Rieske, E., P. Schubert, and G.W. Kreutzberg. 1975. Transfer of radioactive material between electrically coupled neurones of the leech central nervous system. *Brain Res.* **84**:365–382.

Rosenbluth, J. 1973. Postjunctional membrane specialization at cholinergic myoneural junctions in the leech. *J. Comp. Neurol.* **151**:399–405.

Sargent, P.B., K.-W. Yau, and J.G. Nicholls. 1977. Extrasynaptic receptors on cell bodies of neurons in central nervous system of the leech. *J. Neurophysiol.* **40**:446–452.

Scott, S.A. and K.J. Muller. 1980. Synapse regeneration and signals for directed axonal growth in the C.N.S. of the leech. *Dev. Biol.* **80**:345–363.

Staehelin, L.A. 1974. Structure and function of intercellular junctions. *Int. Rev. Cytol.* **39**:191–283.

Stent, G.S., W.B. Kristan, Jr., W.O. Friesen, C.A. Ort, M. Poon, and R.L. Calabrese. 1978. Neuronal generation of the leech swimming movement. *Science* **200**:1348–1357.

Stewart, W.W. 1978. Intracellular marking of neurons with a highly fluorescent naphthalimide dye. *Cell* **14**:741–759.

———. 1981. Lucifer dyes — Highly fluorescent dyes for biological tracing. *Nature* **292**:17–21.

Stuart, A.E. 1970. Physiological and morphological properties of motoneurones in the central nervous system of the leech. *J. Physiol.* **209**:627–646.

Stuart, A.E., A.J. Hudspeth, and Z.W. Hall. 1974. Vital staining of specific monoamine-containing cells in the leech nervous system. *Cell Tissue Res.* **153**:55–61.

Thompson, W.J. and G.S. Stent. 1976a. Neuronal control of heartbeat in the medicinal leech. II. Intersegmental coordination of heart motor neuron activity by heart interneurons. *J. Comp. Physiol.* **111**:281–307.

———. 1976b. Neuronal control of heartbeat in the medicinal leech. III. Synaptic relations of the heart interneurons. *J. Comp. Physiol.* **111**:309–333.

Tisdale, A.D. and Y. Nakajima. 1976. Fine structure of synaptic vesicles in the two types of nerve terminals in crayfish stretch receptor organs: Influence of fixation methods. *J. Comp. Neurol.* **165**:369–386.

Tulsi, R.S. and R.E. Coggeshall. 1971. Neuromuscular junctions on the muscle cells in the central nervous system of the leech, *Hirudo medicinalis*. *J. Comp. Neurol.* **141**:1–16.

Van Essen, D.C. 1973. The contribution of membrane hyperpolarization to adaptation and conduction block in sensory neurones of the leech. *J. Physiol.* **230**:509–534.

Wood, M.R., K.H. Pfenninger, and M.J. Cohen. 1977. Two types of presynaptic configurations in insect central synapses: An ultrastructural analysis. *Brain Res.* **130**:25–45.

Yaksta-Sauerland, B.A. and R.E. Coggeshall. 1973. Neuromuscular junctions in the leech. *J. Comp. Neurol.* **151**:85–100.

Yau, K.-W. 1975. "Receptive fields, geometry and conduction block of sensory cells in the leech central nervous system." Ph.D. thesis, Harvard University, Cambridge, Massachusetts.

———. 1976. Receptive fields, geometry and conduction block of sensory neurones in the C.N.S. of the leech. *J. Physiol.* **263**:513–538.

Zipser, B. 1979a. Identifiable neurons controlling penile eversion in the leech. *J. Neurophysiol.* **42**:455–464.

———. 1979b. Voltage-modulated membrane resistance in coupled leech neurons. *J. Neurophysiol.* **42**:465–475.

7
Neural Circuits Generating Rhythmic Movements

Gunther S. Stent
Department of Molecular Biology
University of California, Berkeley
Berkeley, California 94720

William B. Kristan, Jr.
Department of Biology
University of California, San Diego
La Jolla, California 92093

Rhythmic movements are usually generated by neural elements within the central nervous system (CNS) (Bullock 1961; Delcomyn 1980). In nearly every animal that has been examined, vertebrate or invertebrate, patterns of motor neuron activity closely resembling those driving rhythmic movements in the intact animal continue to be produced in isolated preparations deprived of sensory input. The basic source of such a motor rhythm must therefore be a central rhythm generator composed of elements capable of generating an oscillatory activity pattern. This finding must not, however, be taken to mean that sensory feedback plays no role at all in the realization of rhythmic movements. On the contrary, in most cases the basic rhythm produced by the central generator is subject to influence by sensory feedback, which serves to modulate both the period and the amplitude of the rhythm. In line with this general conclusion, two rhythmic movements carried out by the leech have been found to arise from such central generators—heartbeat and swimming. This chapter summarizes the studies that were carried out to discover how heartbeat and swimming rhythms of the leech are generated (Stent et al. 1978, 1979).

MODELS AND MECHANISMS FOR CENTRAL RHYTHM GENERATION

Models that explain how CNS neurons can generate a rhythmic activity pattern fall into two general classes: (1) endogenous polarization rhythms that depend on an inherent oscillation of the membrane potential of individual neurons and (2) network rhythms that depend on oscillatory activity arising from connections linking a set of neurons with inherently stable membrane potential (Friesen and Stent 1978; Kristan 1980).

Endogenous Polarization Rhythms

Evidence for the existence of neurons capable of endogenous generation of rhythmic impulse bursts in the absence of any rhythmic synaptic input was first provided in molluscan nervous systems (Strumwasser 1967; Alving 1968). The membrane potential of such neurons spontaneously oscillates between a depolarized and a repolarized phase, producing an impulse burst during their depolarized phase. In some cases, the endogenous polarization rhythm has been found to arise from a periodic variation in the intracellular concentration of free Ca^{++} (Meech and Standen 1975; Smith et al. 1975; Gorman et al. 1980). This periodic variation, in turn, arises from the action of two opposing processes, namely, the continuous sequestration, or extrusion, of intracellular Ca^{++} during repolarization and the influx of Ca^{++} from the extracellular medium through voltage-dependent Ca^{++} channels (whose conductance rises with membrane depolarization) to produce depolarization. Moreover, the membrane of the oscillatory cell also contains special K^+ channels, whose conductance rises and falls with the intracellular Ca^{++} concentration. K^+ entry repolarizes the membrane. The period of these cyclical variations in intracellular Ca^{++} concentrations is long (i.e., of the order of hundreds or thousands of milliseconds) compared with the duration of the polarization cycle of action potentials (i.e., of the order of milliseconds).

An impulse train during the depolarized phase is not required for the maintenance of the polarization rhythm (Watanabe et al. 1967; Strumwasser 1971). The cycle period of the endogenous polarization rhythm can be modified by steady passage of current into the cell: Hyperpolarizing current delays, and depolarizing current hastens, the onset of the depolarization phase (Frazier et al. 1967; Kandel et al. 1976). Thus, the possibility exists for physiological control of the period of the endogenous rhythm by setting the level of tonic inhibitory or excitatory input to the endogenously oscillating cell.

Network Rhythms

Self-excitation Networks. One type of network rhythm owes its activity pattern to a neuronal loop consisting of two or more neurons linked by mutually excitatory connections that drive each other to produce impulses at a progressively higher rate. The impulse activity in such a self-excitatory network will oscillate if the network incorporates some restorative feature that terminates impulse production and repolarizes its component neurons as soon as a critical impulse frequency has been attained. In this way, the neurons periodically begin anew their self-excitatory drive to progressively higher impulse frequencies and greater membrane depolarization. Thus, the oscillatory cycle of the cells of self-excitatory networks consists of an active phase of increasing depolarization and impulse frequency and an inactive phase during which the process that stops impulses has repolarized the membrane. Such systems can therefore generate only a single pair of complementary on-off phases of a duty cycle. The mechanisms underlying the restorative termination of impulse production are poorly understood. One proposal is the Ca^{++}-mediated activation of a K^+ conductance increase similar to that previously considered to account for endogenous polarization rhythms (Merickel and Gray 1980). Another possible impulse termination process is the activation of an inhibitory cell (Bradley et al. 1975). A simple realization of such a system consists of three cells—A, B and C—of which A and B are linked by reciprocally excitatory connections and C is provided with excitatory input by cells A and B (Fig. 1a). Cell C, in turn, is linked by inhibitory connections to cells A and B. Cell C has a high threshold for impulse initiation, which is reached only when cell C receives a high level of excitatory input due to high-frequency impulse activity in cells A and B. This circuit generates a rhythm of concurrent impulse bursts in cells A and B, provided that (1) the gain of the feedback loop between cells A and B is greater than unity, (2) the eventual activation of cell C causes hyperpolarization of cells A and B, and (3) cells A and B have a source of tonic excitation to ensure that after hyperpolarization they will resume impulse production and drive each other again to high impulse frequencies. The period of the oscillation depends on such parameters as the impulse activity time of cell C and the time for cells A and B to recover from inhibition. This model has been put forward to account for the generator in the mammalian CNS that drives the contractile rhythm of the diaphragm in breathing (Bradley et al. 1975).

FIGURE 1

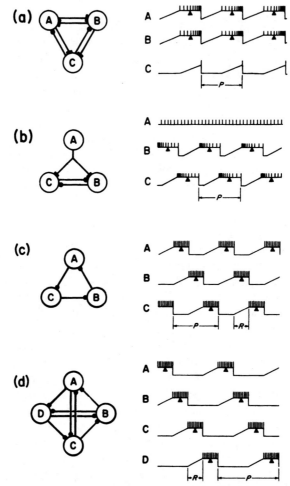

▲ Examples of network rhythms. In this, and all subsequent circuit diagrams, T junctions indicate excitatory and filled circles inhibitory synaptic connections. (*a*) Self-excitation network, in which accelerating impulse production in cells A and B is eventually terminated by activation of the inhibitory cell C. (*b*) Reciprocal inhibition network, in which cells B and C are driven to produce alternating impulse bursts by the tonically active cell A. Cells B and C are subject to adaptation. Three-cell (*c*) and four-cell (*d*) networks with recurrent cyclic inhibition. The traces to the right of the circuit diagrams represent the membrane potential and impulse burst activity in individual cells, as determined by theoretical analysis of the network. (▲) Midpoint, or middle spike, of an impulse burst; *P*, period; *R*, recovery time.

Reciprocal Inhibition Networks. Another type of network rhythm owes its pattern of activity to reciprocally inhibitory loops. In fact, the first proposal for the generation of rhythmic movements is of that type, namely, the "half-center" model (McDougall 1903; Brown 1911). The essential elements of that model are illustrated by the three-neuron network of Figure 1b. Cell A of this network is tonically active, providing excitation to cells B and C, which are connected by reciprocally inhibitory synapses. Cells B and C produce alternating bursts of impulses if they possess some special property such as synaptic fatigue, adaptation, or postinhibitory rebound by which the inhibitory effect of one of these cells on the other decreases as a consequence of past activity (Harmon and Lewis 1966; Perkel and Mulloney 1974).

Recurrent Cyclic Inhibition Networks. In addition to requiring a special property, the reciprocal inhibition network of Figure 1b has one further theoretical limitation as a general rhythm generator: It is limited to the production of biphasic rhythms. Hence, additional network elements are required for the generation of polyphasic rhythms, such as the rhythmic movement of limbs or the contractile waves of serially homologous muscles found in vertebrate and invertebrate locomotion. As Szekely (1967) first realized, introduction of additional elements into the network not only provides for a multiphasic rhythm, but also dispenses with the need for the special property by opening up the possibility for another source of rhythm generation, namely, recurrent cyclic inhibition, as shown in Figure 1c. This network consists of an inhibitory ring formed by three tonically excited neurons, A, B and C, each of which makes inhibitory synaptic contact with, and receives inhibitory synaptic input from, one other cell. If, as indicated in Figure 1c, cell C happens to be in a depolarized impulse-generating state, its postsynaptic cell B must be in a hyperpolarized, inactive state, while its presynaptic cell A is recovering from past inhibition. As soon as cell A has recovered from inhibition and reached its impulse generation threshold, cell C becomes inhibited; this disinhibits cell B and allows it to enter the recovery phase. Once cell B has recovered, it inhibits cell A, thus allowing cell C to begin recovery; and once cell C has recovered, cell B enters its inactive phase and cell A its recovery phase. Thus, one cycle of the oscillation is completed. If the time required for recovery from inhibition of each cell is R, and if the time required for establishing inhibition is small compared with R, then the period of the oscillatory cycle is evidently equal to $3R$. This oscillatory network gives rise to three activity phases separated by phase angles of 120°, and the cycle phases of the three cells progress in a sense opposite to that of the inhibitory connections forming the ring.

Analytical study of a variety of cyclic inhibition networks (Adam 1968; Kling and Szekely 1968) has shown that they produce stable oscillations over a broad range of parameters and can generate as many different phases as the number of cells they contain. Any such ring containing an odd number (N) of cells linked in a manner so that every cell makes inhibitory contact with, and receives inhibition from, one other cell will oscillate with a cycle period $P = NR$. In this N-membered ring, one cell is always in its recovery phase, while the remaining $N - 1$ cells form an alternating sequence of active and inactive phases. By contrast, simple recurrent inhibition rings containing an even number of cells do not oscillate, since they can assume one of two stable states under which either all the even-numbered or all the odd-numbered cells are in the active phase, without any cell being in its recovery phase.

This makes it clear why two-cell networks do not oscillate without a special property. However, with rings containing four or more cells, more complex networks can be formed in which recurrent cyclic inhibition produces an oscillatory activity pattern, even if the number of cells in the ring is even (Kling and Szekely 1968). The simplest such network is a group of four cells each of which makes inhibitory contact with and receives inhibition from two other cells (Fig. 1d). In this network, one cell (say, cell A) is in the active phase, the two cells subject to inhibition by that cell (i.e., cells C and D) are in the inactive phase, and the cell subject to inhibition by these two cells (i.e., cell B) is in the recovery phase. The network gives rise to four phases of activity separated by phase angles of 90°, with the period of the oscillatory cycle being equal to $4R$. That these networks operate as predicted by theory has been confirmed by means of electronic analog circuits (Kling and Szekely 1968; Friesen and Stent 1977).

THE HEARTBEAT

The "heart" of *Hirudo* consists of two contractile lateral vessels, or heart tubes, which form part of a closed circulatory system including also a dorsal vessel and a ventral vessel (which encases the ventral nerve cord) (Mann 1962; see Chapter 3). These longitudinal vessels are joined at the front and rear of the animal to form a continuous conduit extending over the length of the body. In each segment, the major longitudinal vessels are connected via minor, transverse channels and capillaries that irrigate the segmental body tissues. Blood generally flows frontward in the heart tubes and generally rearward in the ventral and dorsal vessels (Boroffka and Hamp 1969). The walls of the heart tubes are ringed by muscles. The periodic contraction of these heart muscles is the heartbeat that provides the pumping action that circulates the blood. The rhythm is segmental, in that the right or left heart-tube section of a particular abdominal segment undergoes its own separate contractile cycle. The period of the segmental heartbeat ranges from about 10 sec at 25°C to 30 sec at 10°C. The pumping action of the heart tubes is coordinated intersegmentally and bilaterally.

The dynamics of the *Hirudo* heartbeat were ascertained by observing the constriction rhythm of the heart tubes in different body segments. For this purpose, the heart tubes were exposed by dissection, and the constrictions of two different ipsilateral or contralateral sections were monitored by visual and electronic means (Thompson and Stent 1976a). The results of a survey of the phase relations of the constriction pattern of the heart-tube sections are presented in Figure 2a. As can be seen, the heartbeat pattern is not bilaterally symmetrical. On the *peristaltic* body side, the segmental heart-tube sections constrict in a rear-to-front sequence, producing a frontward peristalsis. On the *synchronic* body side, the sections constrict almost simultaneously. Even on the peristaltic body side, the peristalsis occurs mainly in the front third of the heart tube, with the sections of the rear half constricting nearly simultaneously. Moreover, the rear sections constrict nearly in the same phase on both sides. Peristaltic and synchronic heartbeat modes are not permanent features of right and left sides; every few minutes the peristaltic side changes to the synchronic mode and the synchronic side changes to the peristaltic mode.

Heart Excitor Motor Neurons

Study of the mechanisms by which the leech CNS generates the heartbeat rhythm became possible with the discovery of excitatory motor neurons called heart excitors (HE cells) (Thompson and Stent 1976a). Paired HE cells that innervate the circular muscles of the ipsilateral segmental heart tube are found in ganglia 3–19 of the nerve cord. In a preparation consisting of a dissected body wall and the heart tube innervated by the ventral nerve cord, HE cells show an activity rhythm in which bursts of action potentials alternate with bursts of inhibitory synaptic potentials (Fig. 2b). The activity cycles of right and left HE cells of the eighth ganglion — cells HE(L,8) and HE(R,8) — are out-of-phase, in accord with the contractile rhythm shown in Figure 2a. Each HE cell action potential is followed, after a constant delay, by an excitatory potential in the muscle fiber of the ipsilateral heart tube, demonstrating that the HE cell is an excitatory motor neuron.

The activity of HE cells of a completely isolated nerve cord preparation is indistinguishable from that of a semi-intact preparation. Thus, the coordinated heartbeat rhythm is produced by a central rhythm generator that does not require sensory input for its patterned output.

By taking simultaneous intracellular recordings from pairs of HE cells in the same ganglion, or from HE cells in different ganglia of the nerve cord, it was found that the HE cell activity cycles are locked into a phase relation that corresponds to that of the segmental heart-tube

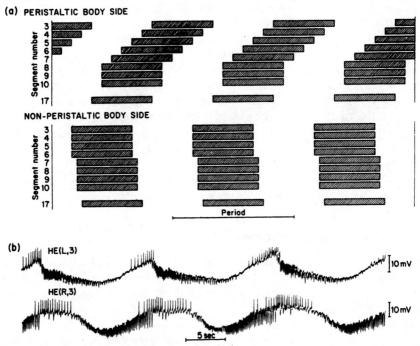

▲ (a) Heart-tube constriction dynamics. Schematic representation of constriction times of segmental heart-tube sections during three heartbeat cycles, illustrating their phase relations on both body sides. Each bar represents the phase angle during which a particular heart-tube section is constricted. On the peristaltic side, there is a rear-to-front progression in constriction of segmental heart-tube sections. On the synchronic (nonperistaltic) side, segmental heart-tube sections constrict nearly in concert along the length of the heart tube. (b) Identification of heart-tube motor neurons (HE cells). Activity rhythm of left (L) and right (R) HE cells of ganglion 3, recorded intracellularly by concurrent microelectrode penetrations of the HE(L,3) and HE(R,3) cell bodies.

constriction pattern shown in Figure 2. Thus, on the peristaltic side the HE cell activity cycles occur in a frontward phase progression, whereas on the synchronic side the activity cycles are nearly in phase. Moreover, a right–left transition in heartbeat coordination mode is reflected in a corresponding phase transition in the HE cell activity cycles. Transitions take place within one to three cycles, either spontaneously (Thompson and Stent 1976b) or in response to stimulation of particular interneurons (Calabrese 1977, 1979a).

Heart Interneurons

Examination of bursts of inhibitory synaptic potentials in HE cells revealed that individual inhibitory potentials recorded in one HE cell are followed with a delay of about 30 msec by an inhibitory potential in another HE cell located in the next posterior ganglion. Hence, the HE cells receive their inhibitory inputs from cells whose axons carry impulses rearward through the connectives. A search for the source of these inhibitory potentials led to the identification of a set of bilateral pairs of rhythmically active cells, designated as heart interneurons (HN cells). Heart interneurons have been identified in ganglia 1–7. The axons of HN cells loop in the ipsilateral neuropil and exit from the ganglion by way of the ipsilateral posterior connective. HN cells send no process into the segmental nerve roots and are therefore "interneurons."

The HN cell activity is rhythmical and consists of an active and an inactive phase, with a period equal to that of the HE cell rhythm, hence the heartbeat period (Fig. 3a). Like the HE cell, the HN cell generates an impulse burst during its active phase and receives an inhibitory synaptic potential burst during its inactive phase. Individual impulses in the HN cell are followed with a constant delay by an inhibitory potential in the HE cell (Fig. 3b), showing that the HN cell impulse burst is the source of the burst of inhibitory synaptic potentials in the HE cell. Upon blocking all inhibitory synaptic input to an HE cell, either by passage of hyperpolarizing current into its presynaptic HN cell(s) or by bathing the preparation in saline containing low Cl^- concentrations (Calabrese 1979b), the HE cell becomes tonically active. Therefore, the activity rhythm of the HE cell is generated by the bursts of inhibitory potentials provided to it by the HN cells.

The manner in which HN cells of ganglia 3, 4, 6, and 7 — i.e., HN(3), HN(4), HN(6), HN(7) — are connected via inhibitory synapses to a series of ipsilateral HE cells in posterior ganglia is shown in Figure 4a. The phase relations of the activity cycles of the HN cells are shown in Figure 4. It is apparent that the contralateral homologs of cells

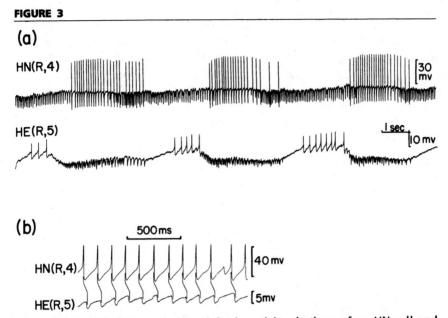

▲ (a) Intracellularly recorded, antiphasic activity rhythms of an HN cell and of an HE cell to which the HN cell provides inhibitory synaptic input. (b) Expansion of a record similar to that drawn in (a). Each action potential in the HN cell is followed with constant delay by an inhibitory synaptic potential in the HE cell.

HN(1), HN(2), HN(3), and HN(4) are active in antiphase, whereas the contralateral homologs HN(6) and HN(7) are active in the same phase. However, of the HN(5) cell pair, only the cell on the synchronic side is rhythmically active, while its contralateral homolog on the peristaltic side is inactive. As far as the longitudinal coordination of ipsilateral HN cells is concerned, the activity cycles of the HN cells that provide synaptic input to the HE cells are active in antiphase on the peristaltic body side, with the activity cycles of HN(3) and HN(4) being 180° out of phase compared with those of HN(6) and HN(7). The cycles of HN(1) and HN(2), which do not contact the HE cells, occur in antiphase with the cycles of HN(3) and HN(4). On the synchronic body side, in contrast, the cycles of HN(3), HN(4), HN(6), and HN(7) occur nearly in the same phase and in antiphase to the cycles of HN(5), HN(1), and HN(2).

During spontaneous transitions from peristaltic to synchronic coordination modes, the phase of the activity cycles of the HN(6) and

FIGURE 4

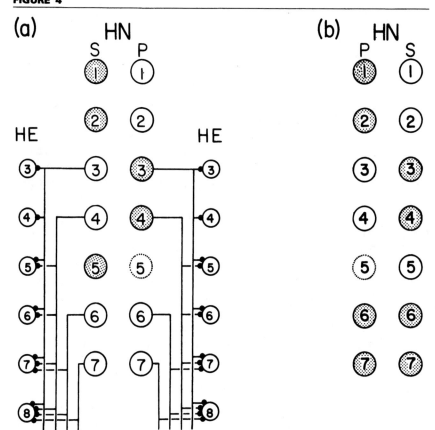

▲ (a) Circuit diagram showing inhibitory connections between HN and HE cells, with left side beating in the synchronous (S) and the right side in the peristaltic (P) constriction mode. A line leaving an HN cell and ending with a solid circle upon HE cell is an identified inhibitory synaptic link (Thompson and Stent 1976c; Calabrese 1977). Cells with matching shading (white or stippled) are active roughly in phase; those with contrasting shading are active in antiphase. The HN(5) cell shown in dotted outline is inactive. The agency which permits only one HN(5) cell to be active is unknown and appears not to involve any of the identified heart interneurons. (b) HN cell activity pattern with the left side beating in the P and the right side in the S mode (Calabrese 1977).

HN(7) cell pairs shifts, while the activity cycle phases of HN(1), HN(2), HN(3), and HN(4) remain invariant. However, the transition is accompanied by the reactivation of the inactive HN(5) on the previously peristaltic side and its resumption of an activity cycle antiphasic to its ipsilateral cell HN(3).

We can now consider how the activity pattern of the HN cells imposes the peristaltic and synchronic activity modes on the HE cells. For this purpose, it may be envisaged that the impulse burst of an HE cell begins (or ends) when the periodic inhibitory input from its presynaptic HN cells falls below (or rises above) the level necessary to maintain the HE cell membrane polarization beyond action potential threshold. Hence, the phase angle that the impulse burst of an HE cell maintains in relation to the heartbeat exceeds by about 180° the average phase angle maintained by the impulse bursts of the HN cells that provide its inhibitory input. For instance, the phase angle of the HE(3) activity cycle would exceed by 180° the average phase angle of the HN(3) activity cycle; the phase angle of the HE(7) activity cycle would exceed by about 180° the average phase angle of the HN(3), HN(4), and HN(6) activity cycles; and the phase angle of the HE(18) cycle would exceed by about 180° the average phase angle of the HN(6) and HN(7) activity cycles. Accordingly, on the peristaltic side, where the activity cycles of the HN cells are out of phase, the HE cell cycles would form a peristaltic phase progression, due, in part, to a differential strength of the influence of different HN cells on the HE cells. For instance, HE(18) is more strongly influenced by HN(7) than by HN(3), whereas HE(4), HE(5), and HE(6) are influenced only by HN(3) and HN(4), but not by HN(6) or HN(7). On the synchronic side, by contrast, where all relevant HN activity cycles occur in nearly the same phase, the HE cell cycles would likewise occur in very nearly the same phase.

The actual phase progression of HE cell activity is governed by three other factors. First, the active phase of HN(3) begins and ends about 40° earlier than that of HN(4); a similar relation exists for the active phases of HN(6) and HN(7). This makes the phase progression more gradual than is implied in Figure 4. The source of this anterior–posterior phase lead is not known. Second, there exists an additional, bilaterally paired HN cell whose presence is known only by its synaptic effects on HE cells and other HN cells. This cell, called HN(X), is active in antiphase to the other HN cells on the synchronic side and is active in phase with HN(3) and HN(4) on the peristaltic side. The third feature that must be taken into account to provide a quantitative explanation of the HE cell activity pattern is an unusual aspect of the inhibitory connections that link HN cells with HE cells. This has been designated presynaptic modulation and arises from two special features of the HN cell synaptic terminals. First, inhibitory synaptic potentials in the HE cell increase in amplitude as the presynaptic HN cell is depolarized (Thompson and Stent 1976c; Nicholls and Wallace 1978; see Chapter 6). Second, each HN cell is linked via rectifying electrical junctions to the axons of anterior HN cells passing through the ganglion. These junctions rectify in such a way that hyperpolarizing current flows from the HN cell located in the ganglion to the axons of other HN cells whose cell bodies are located in anterior ganglia. Depolarizing current flows only in the reverse direction. Thus, variations in membrane potential of an HN cell influence the amplitude of the inhibitory synaptic potentials evoked by its own impulses and also by impulses of the other HN cells to which that HN cell is electrically linked. It appears that the blending of the inhibitory synaptic output of the set of HN cells can provide a reasonably quantitative account of the observed HE cell cycle phase relations on both peristaltic and synchronic sides, based on the known pattern of HN–HE cell synaptic links and on the known relative phase relations of the HN cell activity cycles (Thompson and Stent 1976c).

Source of Interneuron Rhythm

Since generation of the heartbeat depends on the origin of the HN cell activity rhythm, we may inquire into the source of the bursts of inhibitory synaptic potentials in the HN cells during the inactive phase of their activity cycle. As Figure 5 shows, the two HN cells of ganglion 4 are linked by reciprocally inhibitory synapses: Each impulse in one HN cell is followed by an inhibitory synaptic potential in the other. Similar recordings taken from combinations of HN cells show that their inhibitory potentials are accounted for entirely by inhibitory inputs from other HN cells. A network diagram was constructed summarizing all the known connections (Thompson and Stent 1976c; Calabrese 1977; Peterson and Calabrese 1981), and to a first approximation, it accounts for most of the observed phase relations of the HN cell activity cycles shown in Figure 4b. The circuit diagram of Figure 6 shows the essential connections responsible for generating the heartbeat. The reciprocal inhibitory connections between right and left HN(3) and HN(4) pairs explain why the contralateral homologs are active in antiphase. The inhibition by HN(1) and HN(2) of HN(3) and HN(4) on the same side explains why these two sets of HN cells are antiphasically active. The reciprocal inhibition of HN(1) and HN(2) by HN(3) and HN(4) on the same side explains, in turn, why the contralateral HN(1) and HN(2) pairs are antiphasically active (E. Peterson and R. Calabrese, pers. comm.). The inhibition of HN(5) by HN(3) and HN(4) explains why on the synchronic side HN(5) is active in anti-

FIGURE 5

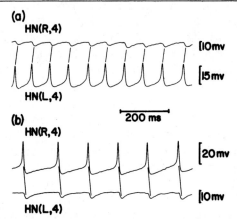

▲ Inhibitory synaptic connections between contralateral HN cell homologs in ganglion 4. Each impulse in one of the homologs is followed with constant delay by an inhibitory synaptic potential in the other homolog.

phase with HN(3) and HN(4), and the bilateral inhibition of HN(6) and HN(7) by the active HN(5) explains why on the synchronic side HN(3), HN(4), HN(6), and HN(7) are all active in the same phase. Finally, this circuit diagram explains why upon right-left transitions in heartbeat coordination mode it is the phase of the HN(6) and HN(7) activity cycle, and not that of the anterior HN cells, that shifts. This transition is attended by a reactivation of the inactive HN(5) on the previously peristaltic side and its resumption of an activity cycle antiphasic to its ipsilateral HN(3), while the active HN(5) on the previously synchronic side becomes inactive. Hence, the in-phase activity cycles of the HN(6) and HN(7) cells on each side of the animal have to shift phase by 180°, as one HN(5) cell is reactivated and the other becomes inactive.

Is the HN cell network the central rhythm generator, or do some other interneurons exist that impose the heartbeat rhythm on the HN cells? One neurophysiological criterion exists for ascertaining whether a rhythmically active neuron is part of a rhythm generator, rather than being merely a follower cell: An evoked transient excitation or inhibition of a generator neuron should shift the phase of the rhythm. According to this criterion the frontmost HN cells qualify as components of the central generator: Passage of depolarizing current into cells HN(1), HN(2), HN(3), or HN(4), causing impulses during their inactive phase, permanently shifts the phase of the heartbeat rhythm (Thompson and Stent 1976b; Peterson and Calabrese 1981).

FIGURE 6

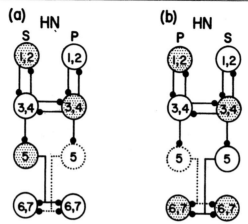

▲ Simplified bilateral circuit diagram of inhibitory connections linking the HN cells. HN cells that are active in nearly the same phase and share the same connectivity pattern are represented as a single circle. (a) Left side beating in the synchronic (S) and right side in the peristaltic (P) mode. (b) Left side beating in the P and right side in the S mode. The inactive cell HN(5) and its synaptic links are shown in dotted outline. (Adapted from Peterson and Calabrese 1981.)

There is indirect evidence that HN cells can produce an endogenous polarization rhythm (Thompson and Stent 1976b; Calabrese 1979b). Certain HN cells continue to be rhythmically active if deprived of inhibitory synaptic input by bathing the nerve cord preparation in Cl^--free saline (Calabrese 1979b). Moreover, the capacity to generate the rhythm is a property not only of the HN cell body but of several independent impulse-initiation sites located in the intersegmental HN cell axons. This follows from the finding that bursts of inhibitory synaptic potentials attributable to an anterior HN cell continue to arise in a posterior HE cell after the connective has been cut between the ganglia in which the HN and HE cell bodies are located (Thompson and Stent 1976c). Moreover, after passing hyperpolarizing current into the cell body of an HN cell in an intact nerve cord, impulse bursts are produced, at nearly normal periods, at an impulse-initiation site located in a posterior ganglion (Thompson and Stent 1976c; Calabrese 1980). It should be noted, however, that there are abundant reciprocal inhibitory connections within the HN cell network that could contribute to the generation of the rhythm in accord with the model shown in Figure 1b.

THE SWIMMING RHYTHM

The swimming movement of leeches is generated by a neuronal network rhythm rather than by an endogenous polarization rhythm (Stent et al. 1978). A leech swims by undulating its extended and flattened body in the dorsoventral plane, forming a wave that travels from head to tail (Figs. 7a and 8a). The moving crests of the body wave are produced by phase-delayed contractile rhythms of the ventral body wall progressing along successive segments and the moving troughs by similar contractions of the dorsal body wall. The forces exerted against the water by these waves propel the leech forward. The period of the segmental contractile rhythm ranges from about 400 msec to about 2000 msec (Gray et al. 1938; Kristan et al. 1974a). The

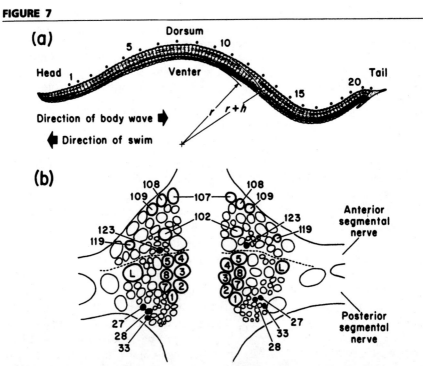

▲ (a) Side view of leech during the swimming movement. The body wave forms a crest with radii of curvature r and $r + h$ in the 8th and a trough in the 16th abdominal segment. (b) Dorsal aspect of a segmental ganglion of *Hirudo* showing the cell bodies of identified motor neurons (heavy outline) and of interneurons (solid black) related to the generation of the swimming rhythm. The cells are numbered according to the system of Ort et al. (1974).

time taken for the travel of the body wave from head to tail is about equal to the period of the contractile rhythm. As a result, at all swimming speeds the animal maintains one full wave length between head and tail (Gray 1968).

In one of the first physiological studies of leech swimming, von Uexkull (1905) decribed the activity pattern of the musculature responsible for these changes in body form. First, the flattening of the body is produced by the tonic contraction of dorsoventral muscles; it optimizes the force that dorsal and ventral body walls exert against the water. Second, the changes in length of the dorsal and ventral body wall segments that form the troughs and the crests of the wave are produced by the phasic local contraction of longitudinal muscles. By exerting a force on the body fluids, the tonic contraction of the dorsoventral muscles generates the antagonistic force necessary for periodic longitudinal distension of the segmental body wall, following relaxation of the segmental longitudinal muscles. Dorsoventral and longitudinal muscles are innervated by motor neurons in the segmental ganglia (Stuart 1970; Ort et al. 1974). These motor neurons are located on the dorsal aspect of the segmental ganglion and numbered as indicated in Figure 7b.

By the late 1930s, the following facts had been established regarding the role of leech nervous system in the generation of the swimming movement. Participation of the head and tail ganglia is not required for production of the body wave, since after decapitation (von Uexkull 1905) or surgical disconnection of these ganglia from the nerve cord, leeches not only still swim but do so more readily, and for longer episodes, than intact animals. Hence, the body wave is generated within the abdominal segments. Moreover, the intersegmental coordination of the contractile rhythm is mediated by the connectives, since after cutting the connectives between any two midbody ganglia, the swimming movements of the body parts anterior and posterior to the cut are no longer phase-locked (Gray et al. 1938). However, the movements of anterior and posterior body remnants do remain coordinated if the connectives are left intact, even if the entire body wall of several midbody segments is removed. Thus, the neuronal activity responsible for coordination of the swimming rhythm travels through ganglia that can neither command contraction of peripheral muscles nor receive sensory input from their own segmental body wall.

The Semi-intact Preparation

This last finding suggested the possibility of developing the semi-intact preparation of *Hirudo* shown in Figure 8b. Here the front and

FIGURE 8

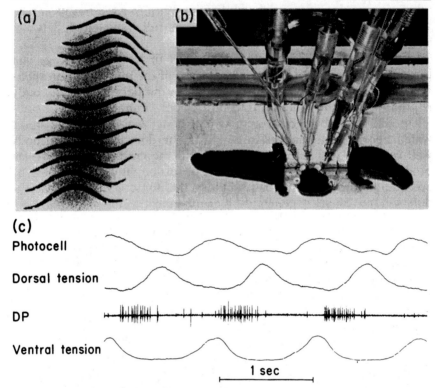

▲ (a) The body wave of a swimming leech, as seen in this composite print of successive frames of a cinematographic record of a free-swimming specimen, with white reference beads attached to the 1st, 5th, 10th, and 15th abdominal segments. The right-to-left horizontal displacement of the animal depicts its true progress in the water. The time occupied by this episode, which corresponds to one cycle period, is about 400 msec. (b) The semi-intact leech preparation pinned venter upward. Front (*right*) and rear (*left*) parts are connected by the exposed ventral nerve cord and its segmental ganglia, of which four are visible. The isolated body wall flap, with sewn-on beads for the attachment of threads leading to tension transducers, is still connected via the left segmental nerves to one of the exposed ganglia. Four glass-tipped suction electrodes for extracellular nerve trunk recordings and two glass capillary microelectrodes for intracellular recordings are shown. (c) Contractile rhythm and segmental nerve activity during a swimming episode of a semi-intact preparation. (*Photocell*) The up and down movements of the head; (*Dorsal tension*), (*Ventral tension*) the output of isometric tension transducers recording the contraction (upward deflection) of the dorsal and ventral sectors of the attached body wall flap; (*DP*) the output of a suction electrode attached to the dorsal branch of the posterior segmental nerve.

rear body remnants carry out coordinated swimming movements. As in all other preparations to be mentioned here, the head and tail ganglia have been disconnected surgically from the nerve cord. At the same time, electrophysiological records are taken from exposed and immobilized parts of the peripheral and central nervous system. Figure 8c shows a record taken from a semi-intact preparation in which the movements were monitored by a photocell registering the interruptions of a light beam caused by up and down movements of the front end. The contractions of the ventral and dorsal sectors of a body wall flap were measured by means of tension transducers, and the efferent impulse activity of the exposed segmental ganglion was monitored by means of a suction electrode attached to the segmental nerve. As can be seen, the dorsal and ventral longitudinal muscles of the exposed body wall flap contract and relax out of phase, with a period of about 1 sec that matches the up-and-down movements of the front end. In the dorsal branch of the segmental nerve, bursts of impulses of uniform amplitude are recorded that are phase-locked with the body wall contractions and the front end movements.

Figure 9a shows another record from a semi-intact preparation during swimming in which an intracellular microelectrode was inserted into a dorsal longitudinal muscle excitor, cell 7, of the exposed ganglion and a suction electrode was attached to the segmental nerve containing the cell 7 axon. In phase with each cycle, the segmental nerve record shows bursts similar to those in Figure 8c, while the intracellular record shows oscillations in the membrane potential of cell 7 that are phase-locked with the impulse burst rhythm. During the depolarized phase of these oscillations, action potential bursts arise that account for the impulse burst rhythm of the segmental nerve branch; during the hyperpolarized phase, cell 7 receives bursts of inhibitory synaptic potentials. Every impulse traveling outward in the axon of cell 7 is followed by an excitatory synaptic potential in the longitudinal muscle fibers in the dorsal territory of the body wall flap. Hence, cell 7 is an excitatory motor neuron of the dorsal longitudinal muscles. Like cell 3 (Fig. 8c), it produces impulse bursts just preceding the contractile phase of the dorsal body wall.

The Swim Motor Neurons

A survey of the cell bodies on the dorsal aspect of segmental ganglia (see Fig. 7b) of semi-intact preparations led to the identification of motor neurons that take part in the generation of the swimming movement (Ort et al. 1974). This ensemble comprises at least seven pairs of excitatory motor neurons to the segmental longitudinal mus-

FIGURE 9

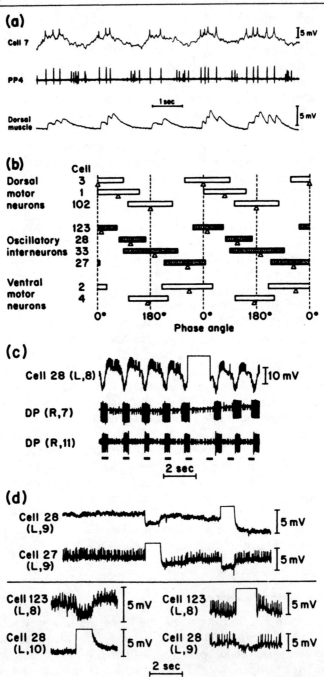

cles, of which four are the dorsal excitors (cells 3, 5, 7, and 107) and three are ventral excitors (cells 4, 8, and 108). Four pairs of motor neurons inhibitory to the segmental longitudinal muscles have been found: Two dorsal inhibitors (cells 1 and 102) cause relaxation of the dorsal body wall and two ventral inhibitors (cells 2 and 119) cause relaxation of the ventral body wall. In addition, a paired excitatory motor neuron cell 109, causes contraction of the dorsoventral (flattener) muscles.

During swimming, cell 109, is depolarized and produces impulses at a high rate, accounting for the flattened body shape and its associated longitudinal stretching force. Meanwhile, the membrane potential of the dorsal and ventral excitors and inhibitors oscillates, just as does that of cell 7 in Figure 9a, between a depolarized and a hyperpolarized phase, with an impulse burst arising during the depolarized phase. Figure 9b summarizes the phase relations of the motor neuron activity observed during the swim cycle. With the phase angle 0° assigned arbitrarily to the middle impulse in the burst of the dorsal excitor cell 3, the impuse burst of the ventral inhibitor cell 119 (not shown in Fig.

◄ (a) Extra- and intracellular recordings from a semi-intact preparation during a swimming episode. (*Upper trace*) Intracellular recording from cell 7 in a midbody ganglion; (*middle trace*) en passant recording from a contralateral nerve (PP4) which contains the axons of cell 7 (class of largest impulses) as well as axons of several other motor neurons; (*lower trace*) intracellular recording from a dorsal longitudinal muscle fiber contralateral to the cell 7 soma, showing excitatory junction potentials time-locked to each cell 7 impulse. (b) Summary diagram of the swimming activity cycles of the oscillatory interneurons and of a representative subset of the motor neurons. Each bar represents the duration, of the impulse burst of the cell, and the triangle marks the burst midpoint. The burst midpoint of cell 3 has been arbitrarily assigned the phase angle 0°. (c) Intracellular microelectrode recording from the left interneuron cell 28 of the 8th abdominal ganglion and suction electrode recordings from the dorsal branch of the right posterior nerve of the 7th [DP (R,7)] and 11th [DP (R,11)] abdominal ganglia during a swim episode of an isolated ventral nerve cord preparation. In this and the following panel, the sharp upward deflection of the intracellular record marks passage into the interneuron of a pulse of depolarizing current. The bars drawn under the bottom trace indicate the times of occurrence of impulse bursts from cell 3 to be expected if passage of current into the interneuron had not shifted the phase of the swimming rhythm. (d) Pairwise intracellular microelectrode recordings from oscillator interneurons within the same ganglion (*upper pair of traces*) or different ganglia (*lower pair of traces*).

9b) occurs at about 0° and the impulse bursts of the ventral excitor cell 4 and of the dorsal inhibitor cell 102 occur at phase angles of about 180°. (The other three dorsal excitors are active in the same phase as cell 3, whereas the other two ventral excitors are active in nearly, but not exactly, the same phase as cell 4). Thus, the dorsal and ventral excitors are active in antiphase, as are the dorsal and ventral inhibitors. However, for the other two inhibitors, the impulse bursts of the dorsal inhibitor cell 1 and of the ventral inhibitor cell 2 occur at the intermediate (albeit mutually antiphasic) phase angles of about 90° and 270°, respectively. The dorsoventral excitor cell 109 produces a continuous train of impulses, although during swimming its membrane is also subject to a low-amplitude polarization cycle that reaches its peak depolarization at a phase angle of about 0°.

Comparison of the activity patterns of homologous swim motor neurons from front to rear along the nerve cord shows that the impulse bursts occur with a progressive phase delay from ganglion to ganglion (Kristan et al. 1974b), in accord with the rearward travel of the body wave (Kristan et al. 1974a; Kristan and Calabrese 1976). Thus, the rhythmic activity of the identified motor neurons in the nerve cord accounts for the basic contractile rhythm of the swimming leech. Accordingly, the two basic questions can be asked: (1) How do the motor neurons of a given ganglion produce the phasic activity pattern in Figure 9b, with cycle periods in the range of 400–2000 msec? and (2) How are the activities of motor neurons of one ganglion coordinated with those of homologous neurons in other ganglia of the cord, so that the rhythms can run from front to rear?

To ascertain whether the answers to these two questions could be found in terms of a network linking the identified motor neurons, a survey was made of the connections that exist between them. Pairs of motor neurons were penetrated with microelectrodes, and current was passed into one cell to observe how it affected the membrane potential of the other. In this way, certain motor neurons were found to be linked by electrical connections that probably serve to synchronize the activity cycle of cells. More important, it was found that the dorsal inhibitors inhibit the dorsal excitors and the ventral inhibitors inhibit the ventral excitors, in addition to their directly inhibitory effect on the longitudinal muscles. Thus, the periodic inhibition provided to each excitor by its homonymous inhibitors constitutes one source of the excitor activity rhythm (Ort et al. 1974). But the source of the activity rhythm of the inhibitors themselves was not accounted for by these findings.

The Oscillatory Interneurons

Understanding of the neuronal control of the swimming rhythm was greatly advanced by the discovery that the motor neurons of an isolated nerve cord, severed from all contact with the leech body, can produce sustained episodes of the swimming activity pattern in response to brief electrical shocks delivered to a segmental nerve (Kristan and Calabrese 1976). Leech swimming, therefore, provides another instance of a locomotory rhythm produced by a central rhythm generator that does not require sensory feedback. Accordingly, a search was carried out in the isolated nerve cord of *Hirudo* for neurons other than the swim motor neurons that might constitute the central swim generator. A neuron was considered to be a candidate component of the generator if, during a swimming episode, (1) its membrane underwent a polarization rhythm that was phase-locked with the impulse burst rhythm of the motor neurons and (2) passage of current into the neuron shifted the phase of impulse burst rhythm of the motor neurons. All the identified motor neurons (with one exception) that met the first of these two criteria failed the second, and, hence, did not qualify as candidate components of the central generator. After an extensive survey of the segmental ganglia, four pairs of neurons in each segmental ganglion were found to meet both criteria (Friesen et al. 1978). They are cells 123, 28, 33, and 27, all having small cell bodies located on the dorsal aspect (Fig. 7b). Figure 9c shows an intracellular recording taken from cell 28 of ganglion 8 during a swimming episode of an isolated preparation consisting of a chain of 18 ganglia. The figure also presents the output of suction electrodes attached to the segmental nerves of ganglia 7 and 11. The initial part of the segmental nerve record shows five cycles of impulse bursts of the dorsal excitor, cell 3, characteristic of the swimming rhythm. Meanwhile, the membrane potential of cell 28 oscillates in a rhythm that is phase-locked with the motor neuron impulse bursts. During its depolarized phase, cell 28 produces an impulse burst whose midpoint occurs at a phase angle of about 90° in the swim cycle of its own segment. Transient passage of depolarizing current into cell 28 of ganglion 8 arrests the impulse burst rhythms of cell 3 in ganglion 7 and 11. After termination of current passage, impulse bursts resume in cell 3, but their phase has been shifted in relation to the swimming rhythm before passage of current. Data similar to those in Figure 9c were obtained also for the other oscillatory interneurons, with the impulse burst midpoint occurring at a phase angle of about 0° for cell 123, 180° for cell 33, and 270° for cell 27.

Anatomical as well as electrophysiological evidence indicates that the oscillatory cells are intersegmental interneurons. First, by staining them with horseradish peroxidase (HRP), cells 28 and 33 can be seen to send an axon into the anterior connective; neither cell sends an axon into the roots of its ganglion. Second, impulses arising in cell 27, cell 28, or cell 33 can be recorded in the connective over a distance of at least five segments to the front, and impulses arising in cell 123 can be recorded over a distance of at least two segments to the rear. No trace of the impulses of any of these four cells can be found in the segmental nerves. Thus, cells 27, 28, and 33 project to more anterior ganglia and cell 123 projects to more posterior ganglia of the cord.

The Oscillator Network

To establish the nature of the network formed by the oscillatory interneurons, intracellular recordings were obtained to ascertain how they are connected within the ganglion and between ganglia. Current was passed into one cell while observing the effects on the other cell's membrane potential or impulse rate. These tests reveal functional connections, but do not distinguish monosynaptic from polysynaptic connections. Figure 9d shows that passage of depolarizing current into either cell 27 or cell 28 of the same ganglion hyperpolarizes the other cell, leading to the inference that cell 27 and cell 28 are reciprocally connected by inhibitory links. In addition, passage of depolarizing current into cell 28 hyperpolarizes cell 123 of a more anterior ganglion and passage of depolarizing current into cell 123 hyperpolarizes cell 28 of a more posterior ganglion. Cells 123 and 28 of different ganglia are, thus, reciprocally connected between ganglia via inhibitory synaptic links. A diagrammatic summary of connections identified in this way is presented in Figure 10. In this diagram, the four interneurons of two ganglia have been placed at the corner of a square, so that their activity phases progress clockwise. As can be seen, within a single ganglion, cells 33, 28, and 123 each make inhibitory connections with that cell that leads it by a phase of 90° in the swim cycle. Furthermore, the antiphasically active cells 27 and 28 are linked via a pair of reciprocally inhibitory connections, and a rectifying electrical junction links cells 33 and 29. Between ganglia, the three interneurons (cells 28, 33, and 27) whose axons project into the anterior connective make inhibitory connections in more anterior ganglia with the same cells with which they also connect in their own ganglion. The one interneuron (cell 123) whose axon projects into the posterior connective makes inhibitory connections in more posterior ganglia with the serial homologs of a cell (cell 28) that follows it by a phase angle of 90° and with which it does not connect in its own ganglion. Although this is not shown in Figure 10, two of the interneurons (cells 28 and 33) have been found to be linked by electrical junctions to their contralateral homologs within the ganglion. Thus, the ganglionic oscillators on the two sides of the animal are coupled.

Output Connections of the Oscillator Network

For the neurons of the identified network to act as components of the central swim oscillator, some cells must make appropriate output connections to the motor neurons. To ascertain the existence of such output connections, intracellular recordings were taken from interneurons and motor neurons (Poon et al. 1978). The results are summarized in Figure 10. As can be seen, the oscillator interneurons are linked to the motor neurons by both inhibitory and excitatory connections, and, in one case, also by a rectifying electrical junction. Some of these connections are intraganglionic. Other connections are inter-

FIGURE 10

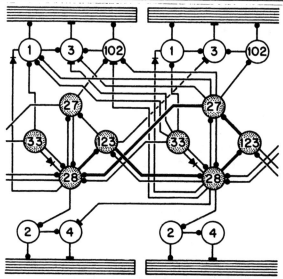

▲ Summary circuit diagram of identified synaptic connections between interneurons (shaded circles), motor neurons (plain circles), and longitudinal muscles responsible for the swimming rhythm. The diode designates a rectifying electrical junction. Connections forming the basic 5-membered, recurrent cyclic inhibition ring are shown as heavy lines.

ganglionic, in the sense that the interneurons whose axons project frontward contact homologous motor neurons in several anterior ganglia and that the rearward projecting interneuron (cell 123) contacts homologs of cell 3 in several posterior ganglia. The output of an electronic model of the network showed that these output connections, in combination with the inhibitory connections of the inhibitory motor neurons to their homonymous excitatory motor neurons, give a reasonably close account of the observed motor neuron duty cycles summarized in Figure 9b (Poon et al. 1978). However, in addition to these four oscillatory interneurons, other cells have been identified as candidate components of the swim generator (Poon 1976; Friesen et al. 1978), but their identification was not pursued. More recent searches resulted in the discovery of the previously unidentified cell 208, an unpaired, segmentally repeated interneuron which possesses a single, posteriorly directed axon and whose membrane potential oscillates during swimming (Weeks 1980). Cell 208 maintains strong excitatory monosynaptic connections with dorsal excitor motor neurons located in posterior segments. In addition, there are inhibitory and excitatory polysynaptic pathways leading from cell 208 to motor neurons located in anterior segments.

Mechanism of the Oscillation

But how do the identified elements of the central oscillator manage to produce the intra- and intersegmentally coordinated swimming rhythm? Analysis of the functional properties of the initially identified interneuronal circuit presented in Figure 10 suggested that it is a complex version of the recurrent cyclic inhibition networks shown in Figure 1c and d (Friesen and Stent 1977). The basic oscillatory circuit of the central swim oscillator would consist of a five-membered ring formed by cells 28 and 123 of an anterior ganglion and cells 28, 123, and 27 of a posterior ganglion. Moreover, the intersegmental inhibitory connections formed by cells 27 and 123 are repeated in several anterior and posterior ganglia, respectively. The basic circuit can be considered as a series of five-membered rings, interconnected from ganglion to ganglion.

To fathom the dynamics of one ring (indicated by the heavy lines in Figure 10), let us suppose that the anterior and posterior cells 28 are in their depolarized state, generating impulses. Then the anterior and posterior cells 123 are inhibited by the two cells 28, and are hyperpolarized and inactive, while the posterior cell 27 is inactive but recovering from inhibition. As soon as cell 27 has recovered and reaches threshold, the anterior cell 28 becomes inhibited. This allows the anterior cell 123 to recover and, once it has recovered, it inhibits the posterior cell 28, allowing the posterior cell 123 to recover. This in turn inhibits the posterior cell 27, allowing recovery of the anterior cell 28. Upon recovery of the anterior cell 28, the anterior cell 123 becomes inhibited, allowing recovery of the posterior cell 28. And upon recovery of the posterior cell 28 and the resulting inhibition of the posterior cell 123, the oscillator has completed one cycle. According to these dynamics, the duty cycles of cell 28 and cell 123 in the same ganglion are nearly antiphasic, whereas the duty cycles of the anterior cells 28 and 123 slightly lead in phase those of their posterior homologs. Thus, as long as the cells are provided with a source of tonic excitation, the basic circuit is capable of producing a crude version of the swimming rhythm, with an essentially two-phase, rather than four-phase, segmental duty cycle. The more complex network of Figure 10 can then be viewed as an elaboration of the basic five-membered intersegmental ring, in the sense that the addition of cell 33 and of reciprocal inhibitory connections between cells 28 and 33 creates a set of subsidiary rings that generate the four-phase segmental duty cycle that is actually observed. The cycle period of this network depends on two parameters: (1) the intersegmental travel time taken by impulses conducted from ganglion to ganglion in the axons of the oscillatory interneurons, and (2) the recovery time taken by each interneuron to reach action potential threshold upon its release from inhibition (Friesen and Stent 1977).

To test the properties of this complex cyclic network, an electronic analog model of the interneurons and their intra- and interganglionic connections was constructed. This model consisted of interconnected electronic "neuromime" elements, which mimic an excitable nerve cell membrane and also provide for the simulation of both excitatory and inhibitory synaptic currents. The analog model circuit produced a good approximation of the observed interneuronal impulse burst relations shown in Figure 9b and gave rise to an appropriate rostrocaudal phase progression of the cycle phases in different ganglia of the nerve cord. This high degree of verisimilitude of the model suggested that the network of Figure 10 constitutes a major component of the central swim oscillator.

Swim-initiating Interneurons

Leeches start to swim in response to stimuli that activate the known mechanosensory neurons, particularly at the rear end of the animal (Kristan et al. 1981). Small water waves that are too weak to activate the touch (T) cells can also evoke swimming (Young et al. 1981), probably via receptors located in the segmental sensilla (Friesen 1981). In isolated nerve cords, the swimming rhythms can be evoked by brief electrical shocks delivered to a segmental nerve and also by addition of serotonin (5-hydroxytryptamine [5-HT]) to the bathing fluid (Weeks 1980; Willard 1981). Since the serotonergic Retzius (R) cells are active throughout swimming episodes, serotonin may be a neurohormone that makes leeches more likely to swim.

Until recently, little or no information was available about the neural pathways that are responsible for the initiation of the swimming rhythm. Whereas the network of Figure 10 could account for the steady-state generation of the rhythm in the nerve cord, it did not provide any explanation for the transition between rest and swimming. This situation changed, however, when a new component of the swim system was discovered in the segmental ganglia, namely cell 204 (Weeks and Kristan 1978). Cell 204 is an unpaired interneuron that sends axons in both the anterior and posterior medial connective (Faivre's nerve) and has a profuse dendritic tree extending over the ganglionic neuropil (Weeks and Kristan 1978; Weeks 1981). Passage of depolarizing current into cell 204 initiates and maintains swimming in semi-intact preparations and in isolated nerve cords (Fig. 11a). Moreover, other stimuli that elicit swimming excite cell 204, and during swimming all cells 204 of the nerve cord are rhythmically active. However, transient passage of current into cell 204 during swimming does not shift the phase of the rhythm and, therefore, cell 204 is not a member of the circuit generating the rhythm. These findings suggest that in the intact leech activation of the set of 204 cells normally initiates a swimming episode.

Another swim-initiating interneuron, cell 205, has been identified (Weeks 1980). Cell 205 is usually present only in ganglion 9, from which cell 204 is usually absent. The anatomy, activity pattern, and synaptic inputs and outputs of cell 205 resemble closely those of cell 204, so that cell 205 may, in fact, be a locally differentiated homolog of cell 204. Cell 205 differs from cell 204 in that it projects its axon only into the anterior medial connective, and, most importantly, in that transient passage of depolarizing current into cell 205, shifts the phase of the swimming rhythm. Moreover, unlike cell 204, cell 205 can receive excitatory input from the S cell via a rectifying electrical junction. Although stimulation of cells 204 and 205 readily initiates swimming, no connection between them and the oscillatory interneurons has been found (J.C. Weeks, in prep.). However, cells 204 and 205 directly excite the recently discovered oscillatory interneuron 208, thereby providing a link from the swim-initiating interneurons to the oscillator.

Since the cell-204 axons course in the medial connective and those of the oscillator interneurons in the lateral connectives (Poon 1976; Friesen et al. 1978), it was possible to examine the functional roles of both types of neurons in initiation, generation, and intersegmental coordination of the rhythm by sectioning the lateral or the medial connectives at a midbody site of an isolated nerve cord (Weeks 1981). According to the proposed mode of operation of the swim oscillator network of Figure 10, the interganglionic oscillator interneuron connections are required for (1) generating the basic motor rhythm, (2) assuring intersegmental coordination of the swim cycles, and (3) setting swim cycle periods of appropriate lengths. Following section in midbody of both lateral connectives in an isolated nerve cord, stimulation of cell 204 anywhere in the cord still elicited the swim rhythm in the entire cord. However, there was no longer any intersegmental coordination of the rhythm across the cut in the connective. By contrast, following section of the medial connective in the midbody, stimulation of any cell 204 still elicited the swim rhythm, which remained coordinated across the site of connective section. Thus, as predicted by the model, oscillator interneurons (and possibly other axons coursing in the lateral connectives) are necessary for intersegmental coordination of the swim cycles.

Another preparation consisted of a single "nearly isolated" ganglion left attached to the anterior and posterior nerve cord only by the medial connectives (Fig. 11b). In response to stimulation of any cell 204, the nearly isolated ganglion generated a swim rhythm with normal cycle periods but slightly abnormal phase relations. That rhythm was, however, completely uncoordinated with the ongoing concurrent rhythm in the anterior and posterior portions of the nerve cord. Thus, contrary to the prediction of the model, the interganglionic oscillator interneuron connections are not required for generating a basic motor rhythm, for which each ganglion appears to contain the necessary elements. It would appear therefore that the swim oscillator network of Figure 10 is incomplete. That is to say, additional connections between identified neurons, additional neurons, such as cell 208, or special properties of previously identified neurons are needed to explain generation of the basic swim rhythm within an "isolated" ganglion.

CONCLUSION

As shown in this chapter, the generation of the two rhythmic movements of leeches — heartbeat and swimming — are now understood at

FIGURE 11

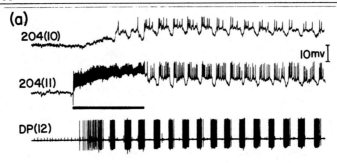

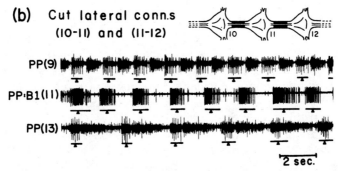

▲ The swim-initiator interneuron cell 204. (a) Intracellular recordings from cell 204 in ganglia 10 and 11 and the DP nerve in ganglion 12 during a swim episode evoked by passage of depolarizing current into cell 204(11) during the 4-sec period indicated by the bar. The swimming rhythm is reflected by the impulse bursts of the dorsal excitor, cell 3, manifest in the DP recording. (b) A nearly isolated ganglion 11 was produced by cutting both lateral connectives on either side of the ganglion (see inset). The traces are simultaneous suction electrode recordings made from the PP nerves of ganglia 9, 11, and 13. The record from PP(11) was taken from the first branch of the PP nerve (PP:B1), where only motor neuron impulse bursts with 180° midpoints are seen. The duration of bursts with midpoints of about 180° is indicated under three cell traces by bars. The recordings were made during a swim episode evoked by passage of depolarizing current into cell 204(10).

a level of detail that is available for few other movements of such complexity. It was possible to gain this understanding largely because of the highly favorable anatomical and physiological properties of the leech nervous system. Of the two movements, the heartbeat turned out to be generated by fewer cells, of which most have probably been identified by now. As for the larger ensemble of cells that constitute the swim-generator circuit, the present state of identification is much less complete. In particular, the elements that permit a nearly isolated single ganglion to generate an approximation to the basic swimming rhythm are still unknown. However, it seems clear that the swim-generator circuit is characterized by a high degree of redundancy: It represents a concatenation of many oscillatory subcircuits, each of which appears to be capable of generating a basic, albeit imperfect, rhythm.

Inasmuch as the two identified neural circuits presented here pertain to the nervous system of the leech, one may ask whether these findings are generally applicable to central nervous oscillators generating rhythmic movements in animals of other species and phyla, particularly in the vertebrates. This question is not easy to answer at this time, because detailed cellular network analyses have thus far been possible only in a very few neurophysiologically favorable preparations. Nevertheless, it is significant that the mechanisms according to which leech heartbeat and leech swimming circuits are thought to generate their oscillations — endogenous rhythmic polarization and recurrent cyclic inhibition — were first proposed to account for generation or rhythmic movements in vertebrate animals. Moreover, the pattern of motor neuron activity in rhythmic movements of vertebrates is not necessarily more complex than the corresponding pattern in analogous movements of leeches. Therefore, the much greater number of neurons in the CNS of vertebrates does not necessarily imply a greater complexity of the central oscillators generating their rhythmic movements; it may only place greater obstacles in the way of identifying the underlying neuronal circuitry. In any case, it is worthy of note that the current list of fundamentally different and theoretically plausible types of neuronal oscillators is not only quite short but also of long standing.

REFERENCES

Adam, A. 1968. Simulation of rhythmic nervous activities. II. Mathematical models for the function of networks with cyclic inhibition. *Kybernetik* **5**:103–109.

Alving, B.O. 1968. Spontaneous activity in isolated somata of *Aplysia* pacemaker neurons. *J. Gen. Physiol.* **45**:29–45.

Boroffka, I. and R. Hamp. 1969. Topographie des Kreislaufsystems und Zirkulation bei *Hirudo medicinalis*. *Z. Morphol. Tiere.* **64**:59–76.

Bradley, G.W., C. von Euler, I. Marttila, and B. Roos. 1975. A model of the central and reflex inhibition of inspiration in the cat. *Biol. Cybern.* **19**:105–116.

Brown, T.G. 1911. The intrinsic factors in the act of progression in the mammal. *Proc. R. Soc. Lond. B* **84**:308–319.

Bullock, T.H. 1961. The origins of patterned nervous discharge. *Behavior* **17**:48–59.

Calabrese, R.L. 1977. The neural control of alternate heartbeat coordination states in the leech, *Hirudo medicinalis*. *J. Comp. Physiol.* **122**:111–143.

―――――. 1979a. Neural generation of the peristaltic and nonperistaltic heartbeat coordination modes of the leech, *Hirudo medicinalis*. *Am. Zool.* **19**:87–102.

―――――. 1979b. The roles of endogenous membrane properties and synaptic interaction in generating the heartbeat rhythm of the leech, *Hirudo medicinalis*. *J. Exp. Biol.* **82**:163–176.

―――――. 1980. Control of impulse-initiation sites in a leech interneuron. *J. Neurophysiol.* **44**:878–896.

Delcomyn, F. 1980. Neural basis of rhythmic behavior in animals. *Science* **210**:492–498.

Frazier, W.T.E., E.R. Kandel, I. Kupfermann, R. Waziri, and R.E. Coggeshall. 1967. Morphological and functional properties of identified neurons in the abdominal ganglion of *Aplysia californica*. *J. Neurophysiol.* **30**:1288–1351.

Friesen, W.O. 1981. Physiology of water motor detection in the medicinal leech. *J. Exp. Biol.* **92**:255–275.

Friesen, W.O. and G. S. Stent. 1977. Generation of a locomotory rhythm by a neural network with recurrent cyclic inhibition. *Biol. Cybern.* **28**:27–40.

―――――. 1978. Neuronal circuits for generating rhythmic movements. *Ann. Rev. Biophys. Biochem.* **7**:37–61.

Friesen, W.O., M. Poon, and G.S. Stent. 1978. Neuronal control of swimming in the medicinal leech. IV. Identification of a network of oscillatory interneurons. *J. Exp. Biol.* **75**:24–43.

Gorman, A.L.F., A. Hermann, and M.V. Thomas. 1980. The neuronal pacemaker cycle. In *Molluscan nerve cells: From biophysics to behavior* (ed. J. Koester and H.J. Byrne), pp. 169–180. Cold Spring Harbor Laboratory, Cold Spring Harbor, New York.

Gray, J. 1968. *Animal locomotion*. Weidenfels and Nicholson, London.

Gray, J., H.W. Lissman, and R.J. Pumphrey. 1938. The mechanisms of locomotion in the leech (*Hirudo medicinalis* Ray). *J. Exp. Biol.* **15**:408–430.

Harmon, L.D. and E.R. Lewis. 1966. Neural modeling. *Physiol. Rev.* **46**:513–591.

Kandel, E.R., T.J. Carew, and J. Koester. 1976. Principles relating to the biophysical properties of neurons and their patterns of interconnections to behavior. In *Electrobiology of nerve synapse and muscle* (ed. J.B. Reuben et al.), pp. 187–215. Raven Press, New York.

Kling, U. and G. Szekely. 1968. Simulation of rhythmic nervous activities. I. Function of networks with cyclic inhibitions. *Kybernetik* **5**:89–103.

Kristan, W.B., Jr. 1980. The generation of rhythmic motor patterns. In *Informa-*

tion processing in the nervous system (ed. H. Pinsker and W.D. Willis), pp. 241–261. Raven Press, New York.

Kristan, W.B. and R.L. Calabrese. 1976. Rhythmic swimming activity in neurons of the isolated nerve cord of the leech. *J. Exp. Biol.* **65**:643–668.

Kristan, W.B., Jr., S.J. McGirr, and G.V. Simpson. 1981. Behavioral and mechanosensory neuron responses to skin stimulation in leeches. *J. Exp. Biol.* (in press).

Kristan, W.B., Jr., G.S. Stent, and C.A. Ort. 1974a. Neuronal control of swimming in the medicinal leech. I. Dynamics of the swimming rhythm. *J. Comp. Physiol.* **94**:97–119.

———. 1974b. Neuronal control of swimming in the medicinal leech. III. Impulse patterns of motor neurons. *J. Comp. Physiol.* **94**:155–176.

Mann, K.H. 1962. *Leeches (Hirudinea). Their structure, physiology, ecology and embryology.* Pergamon Press, New York.

McDougall, W. 1903. The nature of inhibitory processes within the nervous system. *Brian* **26**:153–191.

Meech, R.W. and N.B. Standen. 1975. Potassium activation in *Helix aspersa* under voltage clamp: A component mediated by calcium influx. *J. Physiol.* **249**:211–239.

Merickel, M. and R. Gray. 1980. Investigation of burst generation by the electrically coupled cyberchron network in the snail *Helisoma* using a single electrode voltage clamp. *J. Neurobiol.* **11**:73–102.

Nicholls, J.G. and B.G. Wallace. 1978. Modulation of transmission at an inhibitory synapse in the central nervous system of the leech. *J. Physiol.* **281**:157–170.

Ort, C.A., W.B. Kristan, and G.S. Stent. 1974. Neuronal control of swimming in the medicinal leech. II. Identification and connections of motor neurons. *J. Comp. Physiol.* **94**:121–154.

Perkel, D.H. and B. Mulloney. 1974. Motor pattern production in reciprocally inhibitory neurons exhibiting postinhibitory rebound. *Science* **185**:181–183.

Peterson, E.L. and R.L. Calabrese. 1981. Dynamic analysis of a rhythmic neural circuit in the leech *Hirudo medicinalis*. *J. Neurophysiol.* (in press).

Poon, M. 1976. "A neuronal network generating the swimming rhythm in the leech." Ph.D. thesis, University of California, Berkeley.

Poon, M., W.O. Friesen, and G.S. Stent. 1978. Neuronal control of swimming in the medicinal leech. V. Connections between the oscillatory interneurons and the motor neurons. *J. Exp. Biol.* **75**:43–63.

Smith, T.G., Jr., J.L. Barker, and H. Gainer. 1975. Requirements for bursting pacemaker activity in molluscan neurons. *Nature* **253**:450–452.

Stent, G.S., W.J. Thompson, and R.L. Calabrese. 1979. Neural control of heartbeat in the leech and in some other invertebrates. *Physiol. Rev.* **59**:101–136.

Stent, G.S., W.B. Kristan, Jr., W.O. Friesen, C.A. Ort, M. Poon, and R.L. Calabrese. 1978. Neuronal generation of the leech swimming movement. *Science* **200**:1348–1357.

Strumwasser, F. 1967. Types of information stored in single neurons. In *Invertebrate nervous systems* (ed. C.A.G. Wiersma), pp. 219–319. University

of Chicago Press, Chicago.
──────. 1971. The cellular basis of behavior in *Aplysia*. *J. Psychiatr. Res.* **8**:237–289.

Stuart, A.E. 1970. Physiological and morphological properties of motoneurons in the central nervous system of the leech. *J. Physiol.* **209**:627–646.

Szekely, G. 1967. Development of limb movements: Embryological, physiological and model studies. *Ciba Found. Symp. on growth of the nervous system* (eds. G. Wolstenholme and M. O'Connor), pp. 77–93. Little Brown and Co., Boston.

Thompson, W.J. and G.S. Stent. 1976a. Neuronal control of heartbeat in the medicinal leech. I. Generation of the vascular constriction rhythm. *J. Comp. Physiol.* **111**:261–279.

──────. 1976b. Neuronal control of heartbeat in the medicinal leech. II. Intersegmental coordination of heart motor neuron activity by heart interneurons. *J. Comp. Physiol.* **111**:281–307.

──────. 1976c. Neuronal control of heartbeat in the medicinal leech. III. Synaptic relations of the heart interneurons. *J. Comp. Physiol.* **111**:309–333.

von Uexküll, J. 1905. Studien veber den Tonus III. Die Blutegel. *Z. Biol.* **46**:(N.F. 28):372–402.

Watanabe, A., S. Obara, and T. Akiyama. 1967. Pacemaker potentials for the periodic burst discharge in the heart ganglion of a stomatopod, *Squilla oratoria*. *J. Gen. Physiol.* **50**:839–862.

Weeks, J.C. 1980. "The roles of identified interneurons in initiating and generating the swimming motor pattern of leeches." Ph.D. thesis, University of California, San Diego.

──────. 1981. Neuronal basis of leech swimming: Separation of swim initiation, pattern generation and intersegmental coordination by selective lesions. *J. Neurophysiol.* **45**:698–723.

Weeks, J.C. and W.B. Kristan, Jr. 1978. Initiation, maintenance and modulation of swimming in the medicinal leech by the activity of a single neuron. *J. Exp. Biol.* **77**:71–88.

Willard, A.W. 1981. Effects of serotonin on the generation of the motor program for swimming in the medicinal leech. *J. Neurosci.* (in press).

Young, S.R., R.D. Dedwyler II, and W.O. Friesen. 1981. Responses of the medicinal leech to water waves. *J. Comp. Physiol.* (in press).

8
Neurotransmitter Chemistry

Bruce G. Wallace
Department of Neurobiology
Stanford University School of Medicine
Stanford, California 94305

In spite of the considerable effort devoted in recent years to the physiology, anatomy, and embryology of the leech nervous system, we still know relatively little about the chemistry of leech synapses. This chapter briefly reviews our knowledge of the chemistry of synaptic transmission in the leech and the advantages presented by the leech for neurochemical studies at the cellular level.

In some ways, the turn of this century was a bad time to be a leech. Not only was bloodletting still in vogue, but pioneering neurobiologists were so impressed by the beauty, simplicity, and utility of the leech that bits of its body wall and central nervous system (CNS) were commonplace laboratory material. Just as Retzius and Cajal were fascinated by its anatomy and Whitman by its embryology, early neurochemists, notably Gaskell, were intrigued by the staining properties of cells in leech ganglia. Most striking was the positive chromaffin reaction demonstrated by the Retzius (R) cells and four other small neurons in each segmental ganglion (Gaskell 1914). At that time, the capacity of neurons to reduce the dichromate salts in the chromaffin reaction was attributed to the presence of the catecholamine epinephrine. Gaskell went on to show the similarity of action of epinephrine and nerve stimulation on the peristaltic contractions of the leech lateral heart tubes. He concluded that epinephrine was contained in the cell bodies of R cells, transported along their axons to the periphery, and that the effects of nerve stimulation on the heartbeat were mediated by the release of epinephrine from the terminals of these axons. The similarities in staining properties and effects on the heart between the R cells and neurons in vertebrate sympathetic ganglia and adrenal medulla reinforced Gaskell's belief that some of the effects produced by stimulation of sympathetic neurons were also mediated by the release of epinephrine. He speculated that these peripheral catecholamine-containing systems in vertebrates probably arose during evolution by the migration from the CNS of cells resembling the R cell in the leech. Although many of the details have turned out to be incorrect, Gaskell's speculations remain among the first clear statements of the idea that nerve cells can communicate with their targets by the release of chemical substances. Furthermore, the migration of nerve cells from the neuraxis to form sympathetic ganglia and the adrenal medulla does occur during embryogenesis (Le Douarin and Teillet 1974) and may be relevant to evolution as well.

By providing a reliable and sensitive bioassay preparation, the leech body wall also played a critical role in the experiments that established acetylcholine (ACh) as the transmitter released by neurons innervating vertebrate heart and skeletal muscle. It seems ironic that leech muscle was used to assay ACh for over 50 years before experiments were finally done to identify ACh as the transmitter released by leech motor neurons themselves.

WHY STUDY THE CHEMISTRY OF SYNAPTIC TRANSMISSION?

Individual neurons have distinctive morphologies, form specific synaptic connections, and synthesize, store, release, and respond to particular transmitter substances. To understand how such differences arise, it is necessary to know for each cell the complement and activity of the proteins that produce these neuronal characteristics. Among the distinctive features of neurons that can be quantified in molecular terms are the levels of enzymes involved in transmitter synthesis and degradation. The overall approach is to: (1) assay the levels of activity of enzymes involved in the transmitter metabolism in identified adult neurons, (2) follow the expression of these proteins during development, and (3) determine the changes in transmitter metabolism underlying the modification in signaling between cells in the adult that occur as a result of lesions. For example, by assaying the activity of enzymes during development, one can accurately determine how their levels change in response to changing interactions among cells. Such quantitative results would provide the necessary basis for investigating the mechanisms controlling the differentiation of neurotransmitter function.

WHY STUDY NEUROTRANSMITTER METABOLISM IN THE LEECH?

Many of the advantages of using the leech for the study of neurotransmitter metabolism arise from the simplicity of the leech nervous system and its accessibility for embryological studies (see Chapters 4 and 9). In addition, much is known about the properties of neurons in adult ganglia, how they interact, and how interactions become reestablished and modified following lesions (see Chapters 5, 6, and 10). A feature of the organization of leech ganglia that is important for biochemical studies is that neuronal cell bodies are not sites of synapses; synaptic contacts are made on processes of the cells in the central neuropil (Coggeshall and Fawcett 1964; Muller and McMahan 1976). Thus, simply by cutting the process that connects the soma of an individual neuron to the neuropil, an identified cell can be isolated for biochemical analysis free of contaminating material from other neurons (Sargent 1977). This synaptic organization is typical of many invertebrate ganglia (such as those in the sea hare, *Aplysia*, the snail, and the lobster) and has been exploited by Kandel, Kravitz, Schwartz, McCaman, Osborne, and many others in their studies of transmitter metabolism in individual cells.

A serious disadvantage of the leech for studies of transmitter chemistry is the small size of the neurons; a single cell in *Aplysia*, for example, can be larger than an entire leech ganglion. However, modern radiochemical techniques are sufficiently sensitive for many transmitters and enzymes to be measured accurately even in extracts of single leech cells. Moreover, neurons isolated from the leech CNS (Kuffler and Nicholls 1966; Rude et al. 1969; Sargent 1977; Ready and Nicholls 1979) appear to be significantly less contaminated by nonneuronal cells than those dissected from *Aplysia* or lobster (Giller and Schwartz 1971a; Hildebrand et al. 1974), a feature that may be important in studying the distribution of enzymes involved in transmitter degradation. In addition, the size of neurons varies much less among invertebrate embryos, particularly at early developmental stages when neurotransmitter function may become determined (Goodman et al. 1979).

CHEMICAL ARCHITECTURE OF LEECH SEGMENTAL GANGLIA

5-Hydroxytryptamine-containing Cells in Leech Ganglia

By 1915 it was apparent that the most conspicuous cells in the leech segmental ganglia—the R cells—and four other smaller neurons give a positive chromaffin reaction, and they were assumed to contain epinephrine (Poll and Sommer 1903; Biedl 1910; Gaskell 1914). Over the next 45 years, these original observations were often repeated, but doubts about the identity of the amine arose because of subtle differences between the staining properties of R cells and those of classical chromaffin-positive cells in other tissues (Viallia 1934; Ito 1936; Von der Wense 1939; Perez 1942). With the development of the Falck-Hillarp technique of formaldehyde-induced fluorescence (Falck et al. 1962), it became possible, at least in favorable circumstances, to distinguish among various amines. The yellow fluorescence exhibited by R cells seemed more likely to arise from 5-hydroxytryptamine (5-HT) than from a catecholamine (Bianchi 1967; Kerkut et al. 1967a,b; Ehinger et al. 1968; Kuzmina 1968; Marsden and Kerkut 1969; Rude 1969; Stuart et al. 1974; D. Stuart, pers. comm.). The four other chromaffin-positive neurons also showed 5-HT fluorescence, as did a fifth unpaired cell in each segmental ganglion (Marsden and Kerkut 1969; Rude 1969).

One other neuron that reacts with formaldehyde to give a fluorescent product is found among a cluster of cells at the first branch point of the anterior root (Ehinger et al. 1968; Marsden and Kerkut 1969; Rude 1969; Stuart et al. 1974; D. Stuart, pers. comm.). On the basis of its blue-green fluorescence, this cell apparently contains a catecholamine. Leech ganglia contain dopamine and will synthesize and accumulate radioactively labeled dopamine from labeled tyrosine (McCaman et al. 1973; Sargent 1977). No norepinephrine or epinephrine has been detected in leech ganglia by either of these techniques. Thus, it seems likely that the blue-green fluorescent cells in the anterior roots, as well as those in the first few segmental ganglia, contain dopamine. There are also additional 5-HT-containing cells in ganglia near the head (Marsden and Kerkut 1969; Rude 1969; Stuart et al. 1974). The distribution of cells containing 5-HT and dopamine is illustrated in Figure 1.

This same set of cells can also be stained selectively by incubating ganglia in a dilute solution of Neutral Red dye (Stuart et al. 1974). Neutral Red staining is useful because it does not disrupt the electrophysiological properties of neurons, nor does it interfere with many of the assays used for detecting transmitters and their synthetic enzymes (Stuart et al. 1974; Lent et al. 1979; B.G. Wallace, unpubl.).

Although controversy continues over the interpretation of the flu-

FIGURE 1

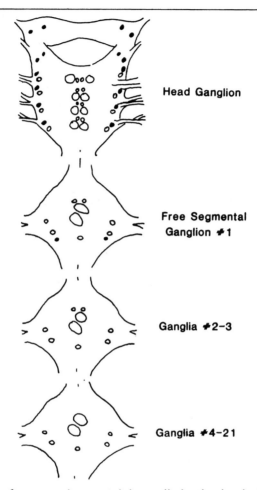

▲ Distribution of monoamine-containing cells in the leech CNS. 5-HT-containing cells (○) are found in each segmental ganglion, the head ganglion, and the tail ganglion (not shown). Dopamine-containing neurons (•) are found in the head and tail ganglia, the first free segmental ganglion, and at the first branch point of the anterior root from each segmental ganglion (not shown). (Adapted from Marsden and Kerkut 1969.)

orescence spectra of amine-containing cells (Bianchi 1974), it has been shown that 5-HT is contained in the R cells and the other neurons that give yellow fluorescence by several independent methods, including paper (Kerkut et al. 1966), ion-exchange (Ehinger et al. 1968), and thin-layer chromatography (Rude et al. 1969; Osborne et al. 1972), quantitative microspectrofluorometry (Ehinger et al. 1968; Rude et al. 1969), and fluorometric (Rude et al. 1969; McCaman et al. 1973), gas chromatographic-mass spectrometric (GC-MS) (McAdoo and Coggeshall 1976), and microradioenzymatic assay of extracts of individual neurons (Lent et al. 1979).

Results from these methods agree in giving an average of 2.3 pmoles 5-HT per R cell in *Hirudo*. This would correspond to an intracellular concentration of approximately 7 mM, if 5-HT were uniformly distributed in the volume of the cell. One of the smaller cells that gives yellow fluorescence, located ventrolaterally in the ganglion, has also been analyzed directly for 5-HT and found to contain 0.2 pmoles 5-HT per cell (Lent et al. 1979). This corresponds to an intracellular concentration of approximately 100 mM, consistent with the observation that these cells display a more intense fluorescence than R cells and also stain more darkly with Neutral Red. Examination by electron microscopy indicates that in leech neurons 5-HT is concentrated in 100-nm dense-core vesicles (see below). It seems reasonable to conclude that the concentration of amine in these vesicles must be greater than 100 nM and that the density of vesicles is considerably higher in the soma of the small ventrolateral cells than in R cells.

Functional Role of R Cells

Cells containing 5-HT are particularly interesting because they appear to perform a variety of functions. Stimulation of R cells produces a decrease in longitudinal muscle tension and accelerates the rate of relaxation following muscle contraction (Mason et al. 1979). Application of 5-HT to leech body wall muscle also reduces muscle tension (Mason et al. 1979) and causes a hyperpolarization that resembles the hyperpolarizing response of muscles to stimulation of cells supplying inhibitory innervation (Walker et al. 1968; Sawada and Coggeshall 1976). Both nerve stimulation and 5-HT application cause an increase in conductance. The potentials become depolarizing if Cl^- is removed from the bathing solution, consistent with the idea that the hyperpolarization arises in both cases from an increase in Cl^- conductance. Lowering external K^+ has little effect on these responses. Two drugs that block the action of 5-HT, bromolysergic acid and cyproheptadine, also block the inhibitory potentials evoked by anterior root stimulation. Examination of leech muscles by electron microscopy has revealed three types of nerve terminals (Yaksta-Sauerland and Coggeshall 1973). One type contains predominantly small, clear vesicles (40–50 nm in diameter), ends directly on muscle cells, and is thought to contain ACh, the excitatory neuromuscular transmitter (see below). The second type of nerve terminal contains 1–10 large, dense-core vesicles (90–120 nm in diameter) and often a few small (50 nm), clear vesicles. These terminals are small and lie close to the muscle membrane, separated from it by a gap of approximately 20 nm. The third type of nerve terminal also contains large, dense-core vesicles, as well as many clear vesicles, and is usually found near clusters of several smaller terminals that contain clear vesicles. Such endings never approach the muscle membrane closely, always being separated from it by either satellite cell processes or clear-vesicle nerve terminals, with which they may form synapses. The 90–120-nm vesicles with eccentric, electron-dense cores found in these terminals are thought to contain 5-HT for three reasons: (1) they resemble dense-core vesicles in R cells that are known to contain 5-HT, (2) their cores reduce dichromate salts (the chromaffin reaction), and (3) the cores can be labeled by injecting the animal with radioactive 5-HT. These observations have led to the suggestion that 5-HT is an inhibitory transmitter for leech body wall muscle (Walker et al. 1968; Sawada and Coggeshall 1976).

On the other hand, recordings from several muscles have failed to reveal inhibitory synaptic potentials that are correlated with action potentials in any of the seven 5-HT-containing cells in the segmental ganglia (Lent 1973; Sawada et al. 1976). Moreover, a neuron that clearly does give rise to inhibitory synaptic potentials in longitudinal body wall muscles does not contain detectable amounts of 5-HT (Sawada et al. 1976). The inhibitory synaptic potentials evoked by stimulation of this cell appear to be monosynaptic, arise from an increase in Cl^- conductance, and are blocked by bromolysergic acid and cyproheptadine, drugs known to block 5-HT receptors. However, the specificity of action of bromolysergic acid and cyproheptadine have not been investigated in leech muscle, nor have the sensitivities of the nerve-evoked and 5-HT potentials to these agents been compared. Thus, one must conclude either that inhibitory neurons do release 5-HT but, unlike all other nerve cells that have been examined, do not contain detectable amounts of the transmitter in the cell soma, or as seems more reasonable, inhibitory neurons release some other transmitter that acts by increasing Cl^- conductance. Another alternative that must be considered is that inhibitory neurons may make strong electrical or chemical synapses onto 5-HT-containing neurons in the periphery, which in turn innervate muscle cells. Therefore, because techniques such as horseradish peroxidase (HRP) injection can be used to identify the peripheral terminals of cells located in leech ganglia (Blackshaw 1981), these ambiguities are amenable to direct experimental analysis.

The distribution of 5-HT terminals in the periphery is characteristic of many amine-containing cells in invertebrates and vertebrates and suggests that transmitter is not directed at a discrete postsynaptic target but rather diffuses over a more widespread area, perhaps interacting with receptors on several different target cells. Compounds acting in this way are often termed neuromodulators and have been implicated in the production of relatively slow, long-lasting modifications of the efficacy of transmission at synapses where other transmitters are released (Kupfermann 1979). In the leech, 5-HT appears to inhibit the release of transmitter from excitatory nerve terminals as well as hyperpolarizing body wall muscles (Walker et al. 1968). These actions may account for much of the effect of R-cell stimulation on muscle tension (Mason et al. 1979) and resemble the effects of amines on neuromuscular transmission in other vertebrate and invertebrate muscles. In the lobster, for example, where glutamate and γ-aminobutyric acid (GABA) have been identified as the excitatory and inhibitory transmitters, 5-HT facilitates the release of transmitter from nerve terminals, and 5-HT, dopamine, and octopamine all produce direct effects on muscle cells (Kravitz et al. 1976). This form of neuromodulation may be important in establishing appropriate muscle tone or maintaining efficient neuromuscular transmission during periods of stress or fatigue (Batelle and Kravitz 1978; Kupfermann 1979; Livingstone et al. 1980).

Another peripheral target for 5-HT released by R cells appears to be the large unicellular glands in the body wall that secrete mucus (Mann 1961). Lent (1973) has shown that the amount of mucus secreted by a patch of body wall increases in proportion to the activity of the R cells. This increase is blocked if transmitter release is prevented by bathing the body wall in saline containing high Mg^{++}, but not if only the ganglion is exposed to Mg^{++}. Direct application of 5-HT to skin causes the release of mucus in a dose-dependent manner, even in the presence of Mg^{++}. Varicosities and terminals containing large, dense-core vesicles are found in the body wall near mucous glands, and occasionally a terminal containing typical dense-core vesicles is seen adjacent to a mucous cell (Yaksta-Sauerland and Coggeshall 1973). Thus, 5-HT released from R-cell terminals in the body wall probably stimulates the mucous glands directly.

5-HT released from R neurons also affects cells in the leech CNS. Micromolar concentrations of 5-HT greatly increase the likelihood that isolated nerve cords will produce episodes of the swim motor program

(Willard 1981), both spontaneously and in response to electrical stimuli. Direct activation of R cells also causes isolated nerve cords to produce the swim motor program, provided the volume of saline bathing the ganglion is sufficiently small. Moreover, leeches normally have a low concentration of 5-HT in their blood, and the level of endogenous 5-HT seems to be correlated with the likelihood that the animal will swim. Elevating the concentration of 5-HT in the blood increases the time spent swimming. The results suggest that 5-HT may modulate the activity of one or more of the central neurons that initiate swimming. This activation may not be mediated by typical synaptic interactions, however. For example, stimulation of R cells or application of 5-HT does not cause any detectable change in the membrane properties of cells 204 and 205, the swim-initiating interneurons; rather, it increases the rate of firing of these cells by increasing the synaptic excitation that they receive (Willard 1981). These effects again resemble those in other invertebrates, where amines apparently interact with particular populations of cells to produce certain types of motor behavior (Berry and Pentreath 1976; Weiss et al. 1978; Livingstone et al. 1980).

Transmitters Synthesized in Leech Ganglia

The idea that R cells release 5-HT as a transmitter is supported by the observation that they synthesize and accumulate radioactively labeled 5-HT when incubated with labeled tryptophan or 5-hydroxytryptophan as a precursor (Hildebrand et al. 1971; Coggeshall 1972; Sargent 1977). Ganglia incubated with labeled choline, tyrosine, tryptophan, and glutamate as transmitter precursors synthesize and accumulate not only 5-HT and dopamine, as expected from the fluorescence studies, but also ACh, GABA, and octopamine (Sargent 1977). The amount of ACh accumulated by ganglia in these experiments exceeds that of any other transmitter by more than two orders of magnitude. Moreover, ganglia contained measurable amounts of endogenous ACh, 5-HT, and dopamine (McCaman et al. 1973; Cammelli et al. 1974). On the other hand, endogenous levels of GABA and octopamine in leech ganglia are below the limits of detection by GC-MS analysis (McAdoo and Coggeshall 1976), suggesting that these transmitters are not likely to be found in more than a few small cells.

Octopamine. The synthesis of octopamine in the leech nervous system is puzzling. An octopamine-containing cell might be expected to stain with Neutral Red, as has been demonstrated in the lobster CNS

(Wallace et al. 1974; Evans et al. 1976). All cells that stain with Neutral Red in the leech, however, appear to contain either 5-HT or dopamine (Stuart et al. 1974). Octopamine is not found in R cells (McAdoo and Coggeshall 1976). It is unlikely to be found in dopamine-containing cells, as the enzyme catalyzing the final step in the biosynthesis of octopamine will also convert dopamine to norepinephrine (Wallace 1976) and no norepinephrine is detected in extracts of leech ganglia nor is any synthesized from labeled precursors (Stuart et al. 1974; McAdoo and Coggeshall 1976; Sargent 1977). Thus, either octopamine is present in one or more of the other 5-HT-containing cells or in cells that fail to stain with Neutral Red.

ACh. The synthesis of relatively large amounts of ACh by leech ganglia suggests that many cells in the leech CNS might release ACh as a transmitter (Sargent 1977). Leech muscle is highly sensitive to ACh, and the contraction evoked by ACh is potentiated by anticholinesterases such as eserine (Fühner 1917). Eserine also potentiates the contractions produced by nerve stimulation, indicating that ACh might be released by leech motor neurons (Bacq and Copée 1937). Recently, microelectrode recordings from single muscle cells have shown that ACh depolarizes leech muscle (Walker et al. 1968) and that the most sensitive regions of the muscle fiber surface correspond to sites of innervation (Kuffler 1978). Excitatory synaptic potentials evoked by nerve stimulation or intracellular stimulation of motor neurons are blocked by curare (Walker et al. 1970; Kuffler 1978), as is the depolarizing effect of ACh (MacIntosh and Perry 1950; Flacke and Yeoh 1968). Two identified motor neurons, the longitudinal motor neuron (L cell) and the annulus erector motor neuron (AE cell) synthesize and accumulate [^{3}H]ACh from [^{3}H]choline, and extracts of these cells have been shown to contain the enzyme that catalyzes this reaction, choline acetylase (Sargent 1977). Thus, there is considerable pharmacological and biochemical evidence that ACh is an excitatory neuromuscular transmitter in the leech.

The finding that the L and AE motor neurons are cholinergic raises two questions about the biochemical architecture of leech ganglia: Do all excitatory motor neurons release ACh, and are motor neurons the only cholinergic cells in the ganglion? These questions can be approached by measuring the distribution of choline acetylase activity, which seems to be a reliable indicator of cholinergic function (Sargent 1977). Twelve excitatory motor neurons analyzed so far contain levels of choline acetylase similar to those found in the AE and L motor neurons (Table 1). One other identified cell that has a comparable level

TABLE 1

Choline acetylase in identified cells

Cell number	Cell function	N	Activity (pmoles/cell/hr)
Identified excitatory motor neurons			
3	dorsomedial longitudinal	4	0.83 ± 0.22
4	ventromedial longitudinal	11	0.54 ± 0.14
5	dorsolateral longitudinal	3	1.84 ± 0.51
7	dorsal longitudinal	3	2.14 ± 0.57
8	ventrolateral longitudinal	2	3.92 (1.66, 6.18)
107	dorsomedial longitudinal	3	2.74 ± 0.73
106	lateral longitudinal	2	4.12 (4.58, 3.66)
11	circular (ventral)	2	0.61 (0.76, 0.47)
12	circular (ventral)	2	1.65 (1.05, 2.25)
112	circular (dorsal)	1	4.05
110	oblique	1	2.98
AE	annulus erector	9	1.09 ± 0.25
L	large longitudinal	8	2.41 ± 0.46
HE	heart excitor	8	0.93 ± 0.14
Identified inhibitory neurons			
1	dorsal longitudinal inhibitor	1	0.0
2	ventral longitudinal inhibitor	3	0.0
101	flattener inhibitor	2	0.0
102	dorsal inhibitor	1	0.0
Other neurons			
	anterior pagoda	5	2.21 ± 0.34
	nut	5	1.25 ± 0.42
	N mechanosensory	7	0.0
	Leydig	3	0.0
	middle cell	2	0.0

Data are expressed as mean ± s.e.m.; N gives the number of determinations. Individual cells were isolated using nylon loops (Wallace 1981) and assayed as described by Sargent (1977).

of choline acetylase is the heart excitor motor neuron (HE cell), which also synthesizes [^{3}H]ACh from [^{3}H]choline and accumulates it (Fig. 2). Two other neurons that contain choline acetylase are the anterior pagoda, the largest neuron on the ventral surface of the anteriolateral packet, and the nut, a neuron just anterior to the AE cells (Sargent 1975). The functions of these cells are not known. Although they send processes through the roots to the periphery, they do not appear to innervate body wall muscles.

FIGURE 2

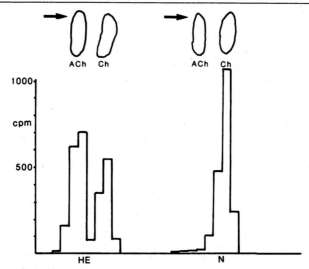

▲ Synthesis of ACh by HE motor neurons. Ganglia were incubated in L-15 medium containing 20 μM [³H]choline (10 Ci/mmole) for 5 hr, rinsed, and eight HE motor neurons and eight N mechanosensory cells were isolated, pooled, and analyzed for synthesis of [³H]ACh by paper electrophoresis (Sargent 1977). Both HE and N cells contained [³H]choline. In addition, HE cells synthesized and accumulated 28.6 fmoles/cell of [³H]ACh; no [³H]ACh was detected in the N-cell extract. In five experiments, HE cells synthesized and accumulated an average of 28.8 ± 2.2 fmoles ACh/cell (mean ± S.E.M.).

Other transmitters. Little information is available about the transmitters released by other cells in the leech nervous system. Mechanosensory cells, for example, do not synthesize ACh, GABA, or any of the monoamine transmitters (Sargent 1977). From his studies on the synthesis of transmitters by intact ganglia, Sargent (1977) concluded that only one-fourth of the 350 – 400 neurons in a leech ganglion are likely to release ACh, 5-HT, dopamine, octopamine, or GABA. This leaves two obvious categories of transmitters to be investigated, neuropeptides and amino acids such as glutamate and glycine. Although direct techniques for identifying neurons releasing amino acids need to be developed, immunological methods can be exploited in the study of neuropeptides (Hökfelt et al. 1980). The first application of this technology to the leech revealed the presence in the CNS of cells and nerve terminals that stain with antiserum against Leu-enkephalin (Zipser 1980). The possibility that other neuropeptides originally discovered in vertebrates are present in specific neurons in the leech remains to be investigated.

ENZYMES OF TRANSMITTER SYNTHESIS AND STORAGE

To identify the transmitter released by a particular neuron, assaying the activity of enzymes catalyzing neurotransmitter synthesis is often more sensitive than measuring the level of the transmitter itself. It has been a general finding that the presence of a synthetic enzyme is a reliable indicator of the transmitter released (Giller and Schwartz 1971a; Hildebrand et al. 1974). This generalization also seems to hold for cells in the leech nervous system. The enzyme catalyzing the final step in 5-HT synthesis, aromatic amino acid, has only been detected in extracts of 5-HT-containing cells (Coggeshall et al. 1972). On the basis of the limits of detectability, the amount of enzyme in a single R cell is at least 870 times that in a mechanosensory cell or motor neuron (Table 2). A similar divergence is found in the distribution of choline acetylase. Whereas excitatory motor neurons contain choline acetylase

TABLE 2

Aromatic amino acid decarboxylase activity in identified cells

	Number of cells in sample	Activity (pmoles/cell/hr)[a]	
R cells	2–4	46.10 ± 3.29	(28)
AE motor neurons	4–8	0.0	(4)
N sensory neurons	4–6	0.0	(3)
P sensory neurons	6	0.0	(1)
Buffer	—	0.42 ± 0.05	(17)

Cells were isolated using nylon loops (Wallace 1981) and extracted in 5 μl of 50 mM Na phosphate (pH 6.5), 0.1% (v/v) Triton by freezing and thawing. Reaction was begun by adding 1 μl of substrate mix to give, in final concentrations: 0.2 mM [^{3}H]DOPA (1 Ci/mmole, Amersham), 6 μM pyridoxal phosphate, 0.1 mM 2-aminoethylisothiouronium bromide, 0.1% Triton (v/v), 50 mM Na phosphate (pH 6.5). Samples were incubated 60 min at 38°C, the reaction stopped with 1 μl of concentrated (×7) electrophoresis buffer containing unlabeled standards, and placed on ice. Samples were spun, 6 μl removed, and DOPA and dopamine separated by high-voltage paper electrophoresis at pH 2 (Hildebrand et al. 1971). Standards were visualized with ethylenediamine spray (Schneider and Gillis 1965) and the dopamine spots cut out, eluted, and counted in 5 ml of Protosol fluor. Recovery, determined by including [^{3}H]dopamine as an internal standard, was approximately 60%.

[a] Activity is expressed as mean ± s.e.m.; the number of samples analyzed is given in parentheses. The limit of detectability, taken as twice background, corresponds to 0.42 pmoles/assay/hr.

(~2 pmoles/cell/hr), no activity can be detected in extracts of mechanosensory or R cells. This corresponds to a ratio of motor to sensory or R-cell activities of at least 100:1 (Sargent 1977). Clearly, one important step at which control is exerted during the differentiation of neurons is the expression of enzymes catalyzing the synthesis of neurotransmitters.

In one study, choline acetylase activity was found in extracts of R cells (Coggeshall et al. 1972), but this has not been confirmed using a more sensitive assay (Sargent 1977). Although the reason for the discrepancy is not clear, it is worth noting that small cells often are associated with the stalks of the R cell and may be included, inadvertently, along with the R cells during dissection (Fuchs et al. 1981). On the other hand, the possibility remains that the R cell may contain two different transmitter compounds; more than one transmitter has been detected in the cell body of other invertebrate neurons (Brownstein et al. 1974; Cottrell 1974, 1976; Hanley et al. 1974). Recent studies combining formaldehyde-induced fluorescence and immunohistochemical techniques in the vertebrate CNS indicate that the joint presence of an amine and a neuropeptide in the same cell may be common (Hökfelt et al. 1980).

As for enzymes involved in transmitter degradation, studies in the lobster and in *Aplysia* have indicated that the levels of degradated enzymes are similar in all cells whether or not they release a particular transmitter (Hall et al. 1970; Giller and Schwartz 1971b; Hildebrand et al. 1974). These results suggest that degradative enzymes may be thought of as constitutive neuronal components. This may be true for some general catabolic enzymes such as monoamine oxidase, which degrades dopamine, octopamine, and 5-HT and is a common constituent of mitochondrial membranes (Schnaitman et al. 1967). However, can it be expected that cells have comparable levels of an enzyme such as acetylcholinesterase (AChE), whose only apparent physiological function is to hydrolyze ACh? AChE is, of course, not uniformly distributed in the nervous system but is concentrated at synapses where ACh is released. This accumulation of AChE generally marks the postsynaptic cell at cholinergic synapses but often is not useful in identifying the presynaptic terminal or the cell body of a neuron releasing ACh.

AChE also plays a role within the nerve terminals and cell bodies of cholinergic neurons, where it is thought to limit the accumulation of ACh (Birks and MacIntosh 1961; Potter 1970; Sargent 1977). Although such an intracellular function might only be required in neurons that release ACh, it was found upon isolation of individual cells or axons from the lobster and *Aplysia* nervous systems that AChE activity is associated with neurons releasing ACh as well as with neurons releasing other transmitters (Giller and Schwartz 1971b; Hildebrand et al. 1974). Much of the cholinesterase (ChE) activity measured in these studies, however, may have been nonneuronal, that is, present on or in glial and connective tissue cells contaminating the samples (Giller and Schwartz 1971a,b; Hildebrand et al. 1974). Individual leech neuronal cell bodies can be isolated relatively free of such contaminants (Kuffler and Nicholls 1966; Rude et al. 1969; Sargent 1977), and therefore the distribution of AChE has been measured in cholinergic and noncholinergic neurons in leech ganglia. Intracellular AChE activity was determined by pretreatment of the ganglia with the irreversible inhibitor echothiophate, which blocks ChE in the extracellular space (Brimijoin et al. 1978). It was found that cells releasing ACh as a transmitter have approximately tenfold higher levels of intracellular AChE activity, whereas all neurons have comparable amounts of activity associated with the cell surface (Wallace 1981). These results suggest that there may be three pools of AChE. All neurons would have similar levels of surface enzyme and a small amount of intracellular ChE. Cholinergic neurons would be unique in having an additional pool of intracellular enzyme that regulates the accumulation of ACh. Expression of this particular pool of intracellular AChE would be another step in neuronal differentiation.

Different molecular forms of AChE have been identified in vertebrate tissue (Bon et al. 1979; Massoulié et al. 1980). In the rat, a high-molecular-weight form seems to be localized specifically at the neuromuscular junction (Hall 1973). Experiments have failed to reveal any difference in the molecular properties of intracellular and extracellular enzyme in the leech CNS (B. Wallace, unpubl.). Enzyme extracted from leech ganglia in phosphate-buffered 0.5% Triton sediments on sucrose gradient velocity sedimentation primarily as a single peak corresponding to an $S_{20,w} = 6.5$. There is also a small peak of activity at 4.5S. The pattern of activity on the gradient is the same if the ganglia are pretreated with echothiophate, although the total activity is reduced by approximately 97%. These results indicate that, once solubilized in Triton, there is no detectable difference in the molecular size of intracellular and surface AChE.

There is also no significant heterogeneity in the catalytic properties of AChE isolated from leech ganglia. Enzyme from the leech CNS resembles "true" AChE. It is inhibited over 90% by 10^{-6} M BW284c51, a selective AChE inhibitor, and is relatively insensitive to inhibition by tetraisopropylpyrophosphoramide (ISO-OMPA), an inhibitor of vertebrate "pseudo" or butyrylcholinesterase. However, the substrate specificity of ChE from the leech nervous system is broader than that of vertebrate AChE. For the leech enzyme, the kinetics of hydrolysis of butyrylthiocholine and acetylthiocholine are approximately the same. Leech blood also has ChE activity; it resembles vertebrate butyrylcholinesterase in its susceptibility to inhibition by ISO-OMPA. All activity in extracts of isolated cells is inhibited by BW284c51 and insensitive to inhibition by ISO-OMPA. Thus, there are no obvious differences in the molecular or catalytic properties of AChE located within cholinergic cells and the enzyme associated with other neurons.

DISTRIBUTION AND PROPERTIES OF NEUROTRANSMITTER RECEPTORS

Leech neurons with different functions can be distinguished by their sensitivity to compounds applied iontophoretically to the cell body (Sargent et al. 1977). The touch (T), pressure (P), and nociceptive (N) mechanosensory neurons and L and AE motor neurons all respond to ACh. The P and N cells are depolarized, and the T and L cells give mixed depolarizing and hyperpolarizing responses, whereas the AE cell is hyperpolarized. Some of these same cells respond to 5-HT, GABA, glycine, or dopamine, whereas none of these cells responds to glutamate. For example, only the P cells and the medial N cell are sensitive to dopamine. For these studies, fine pipettes and low doses were used so that the responses elicited were derived from receptors on the surface of the cell body and not from within the neuropil, as shown by the short latency and rapid time to the peak of the responses and their sensitivity to very small movements (a few micrometers) of the iontophoretic pipette.

Because no synapses are present on the soma of neurons in the leech CNS (Coggeshall and Fawcett 1964; but see Muller et al. 1978), receptors for neurotransmitters on the soma must be extrasynaptic. Extrasynaptic receptors are found on neurons in other invertebrates (Gerschenfeld 1973) but have been best characterized in vertebrate skeletal muscle (Fambrough 1979). The function of extrasynaptic receptors and their relationship to subsynaptic receptors is unknown, although in *Aplysia*, for example, subsynaptic and extrasynaptic receptors can have similar ionic and pharmacological properties (Blankenship et al. 1971; Kehoe 1972). In leech ganglia, on the other hand, there can be clear differences in the pharmacology of extra- and subsynaptic receptors, and the effects produced by application of a particular transmitter can differ depending on whether it is applied to the soma or to the "stalk" of the neuron (Sargent et al. 1977). This makes it unlikely that extrasynaptic receptors on leech neurons represent incomplete localization of receptors to the subsynaptic surface or an intermediate stage in the transport of receptors from the cell body to the subsynaptic membrane.

One serious complication with iontophoretic mapping of receptor distribution is the specificity of the stimulus. When current is pulsed through an iontophoretic pipette, changes in tonicity, ionic strength, and pH can occur in the immediate vicinity of the pipette tip. Such changes can evoke an electrical response in the cell (MacDonald et al. 1979). For example, solutions containing 5-HT for iontophoresis are often acidified to pH 3 or 4 to reduce the rate of 5-HT oxidation. An unexpected finding is that cells in leech ganglia respond to pulses from a pipette filled only with leech saline adjusted to pH 3 (L. Henderson, pers. comm.). These responses are very similar to those elicited by 5-HT itself — depolarization of the lateral N mechanosensory cell and hyperpolarization of the AE motor neuron. Similar effects are produced by pressure application of acidified saline, indicating that changes in pH alone can evoke different responses in different cells. One method that may prove useful for investigating neurotransmitter receptors is application of neutralized, isotonic solutions of the drug in saline using a pressure system similar to that used routinely for intracellular injection and modified with an electronic valve to give pressure pulses of controlled duration (MacDonald et al. 1979; L. Henderson, pers. comm.).

Bath application is clearly less desirable as an alternative method of drug delivery. In the leech, as in other invertebrates, cells can have different receptors for the same agent on different parts of the cell surface (Gerschenfeld and Paupardin-Tritsch 1974; Sargent et al. 1977). Although different receptors can be so intimately intermingled as to be unresolvable even with highly localized iontophoretic application, the likelihood of activating a variety of different extra- and subsynaptic receptors is increased when drugs are applied in the bathing solution. Moreover, the time resolution of response will be reduced.

A preparation that promises to be especially useful in comparing extra- and subsynaptic receptors on leech neurons is the culture of individual, identified cells (Ready and Nicholls 1979; Fuchs et al. 1981). Such isolated cells survive, extend elaborate branches, and establish connections in vitro. Because an identified cell can be seen in its entirety, responses can be elicited by iontophoretic or pressure application over its entire arborization. The same cell can then be examined in the electron microscope, and each location at which a response is recorded can be positively identified as synaptic or extrasynaptic. Experiments on single cells in culture have demonstrated that similar, apparently extrasynaptic, receptors can occur not only on the soma but also along highly branched neurites (Henderson 1981).

One possible function mediated by extrasynaptic receptors is the response of neurons to neuromodulators and hormones. As described above, the concentration of 5-HT in the blood can influence the likelihood that a leech will swim (Willard 1981). Similar changes in the pattern of activity of neurons in response to changes in the levels of 5-HT and octopamine in the circulation have been reported in the lobster (Livingstone et al. 1980). Substances in the circulation might modify the activity of a group of neurons in the CNS through interaction with specific extrasynaptic receptors. Unlike synaptic transmission, in which specificity is achieved in part anatomically, specificity in the response to circulating factors can be achieved only by the distribution and nature of the receptors themselves.

FIGURE 3

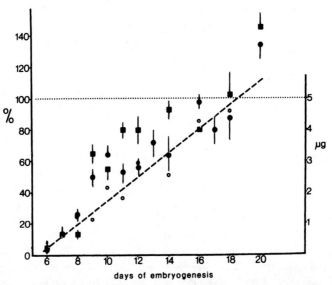

▲ Appearance of transmitter synthetic enzymes. Specific activities of choline acetylase (●) and aromatic amino acid decarboxylase (■) as percentages of the adult values and total protein (o, ---------) in embryonic central nerve cord are plotted against developmental age. Choline acetylase was assayed as described by Sargent (1977); aromatic amino acid decarboxylase assay is described in the note to Table 2. Error bars are ±S.E.M. (N = 2–7).

FIGURE 4

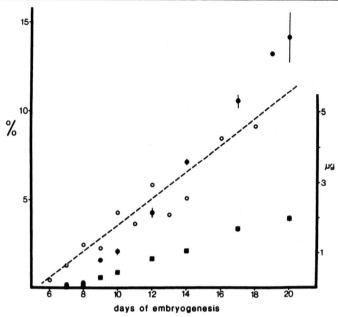

▲ Synthesis and accumulation of neurotransmitters by embryonic ventral nerve cord in vitro. Ventral nerve cord was isolated from embryos at each stage, incubated in L-15 medium with [^{3}H]choline or 5-[^{3}H]HT for 4 hr, washed, extracted, and analyzed for ACh (■) or 5-HT (●) by high-voltage paper electrophoresis (Hildebrand et al. 1971; Sargent 1977). Error bars, shown where they exceed the size of the spot, are ± S.E.M. (N = 2–4). Synthesis and accumulation are expressed as percentages of levels found in adult cords incubated under the same conditions. (○, ---------) Total protein in embryonic cord.

NEUROTRANSMITTER METABOLISM DURING DEVELOPMENT

Relatively little information is available about the acquisition of neurotransmitter function during the development of the leech nervous system. Studies in *Hirudo* have shown that biosynthetic enzymes for two transmitters, choline acetylase and aromatic amino acid, can be detected in extracts of embryonic ventral nerve cords. The specific activity of both enzymes increases sharply early in the second week of embryogenesis, at about the time that segmental ganglia first become recognizable, and reaches and exceeds adult levels by 21 days, when embryonic development is complete (Fig. 3). Early in the second week,

TABLE 3

Aromatic amino acid decarboxylase activity in R cells

Stage	pmoles/cell/pl	pmoles/pliter/hr
13	0.22, 0.22	0.05
14	0.42 ± 0.13 (S.E.M., $N = 5$)	0.08
18	0.91	0.13
Adult	47.9 ± 3.0 (S.E.M., $N = 6$)	0.09

Cells were isolated and analyzed as described in the note to Table 2.

neurons also begin to acquire the capacity to synthesize and accumulate 5-HT and ACh from radioactively labeled precursors, indicating that other proteins involved in the synthesis and storage of transmitters are also present at this time (Fig. 4). In *Hirudo*, there is insufficient information on the morphology and physiology of embryonic neurons to correlate these biochemical changes with other developmental events.

Despite the small size of embryonic neurons, it is technically feasible to isolate cells from developing ganglia and assay their complement of enzymes. For example, individual R cells have been identified and isolated from *Hirudo* embryos at various ages and the amount of aromatic amino acid activity measured. Corrected for differences in cell volume, the concentration of enzyme in cells from 13-, 14-, and 18-day embryos is comparable to that found in adult R cells (Table 3). This suggests that neurochemical differentiation may be an all-or-none phenomenon, and that in response to some developmental signal R cells may rapidly begin synthesizing and accumulating the enzymes for 5-HT synthesis at the same levels found in the adult. The 200-fold increase in total activity per cell that occurs between embryonic day 13 and adulthood would then not represent a specific response by this enzyme system but rather would result from the change in the size of the cell soma.

CONCLUSIONS

One way that neurons communicate with their targets is through the release of a chemical neurotransmitter or neuromodulator. Whereas considerable information regarding the release of neurotransmitters can be surmised from electrophysiological recordings, knowledge of the chemistry of synapses is essential to understand how the expression of neurotransmitter phenotype is regulated during development and how long-term changes in signaling in the nervous system are produced. The nervous system of the leech offers the opportunity to study the neurochemistry of individual, identified cells during embryogenesis, in the adult, and in ganglia and isolated neurons maintained in culture. Correlation of changes in the physiology, morphology, and chemistry of individual cells during development and in response to lesions of embryonic or adult ganglia should provide valuable insights into the mechanisms that determine transmitter function, as well as other, less readily quantified features that distinguish neurons from one another.

REFERENCES

Bacq, Z.M. and G. Coppée. 1937. Action de l'ésérine sur la préparation neuromusculaire du siponcle et de la sangsue. *C. R. Soc. Biol.* **124**:1244–1247.

Batelle, B.-A. and E.A. Kravitz. 1978. Targets of octopamine action in the lobster: Cyclic nucleotide changes and physiological effects in haemolymph, heart and exoskeletal muscle. *J. Pharmacol. Exp. Ther.* **205**:438–448.

Berry, M.S. and V.W. Pentreath. 1976. Properties of a symmetric pair of serotonin-containing neurones in the cerebral ganglia of *Planorbis*. *J. Exp. Biol.* **65**:361–380.

Bianchi, S. 1967. On the different types of fluorescent neurons in the leech (*Hirudo medicinalis*). *Atti. Soc. Peloritana. Sci. Fis. Mat. Nat.* **13**:39–47.

―――――. 1974. The histochemistry of the biogenic amines in the central nervous system of *Hirudo medicinalis*. *Gen. Comp. Endocrinol.* **22**:245–249.

Biedl, A. 1910. Ueber das Adrenalgewebe bei Wirbellosen. *8th Int. Congr. Zool. Graz.* pp. 503–505.

Birks, R.I. and F.C. MacIntosh. 1961. Acetylcholine metabolism of a sympathetic ganglion. *Can. J. Biochem. Physiol.* **39**:787–827.

Blackshaw, S. 1981. Morphology and distribution of touch cell terminals in the skin of the leech. *J. Physiol.* (in press).

Blankenship, J.E., H. Wachtel, and E.R. Kandel. 1971. Ionic mechanisms of excitatory, inhibitory, and dual synaptic actions mediated by an identified interneuron in abdominal ganglion of *Aplysia*. *J. Neurophysiol.* **34**:76–92.

Bon, S., M. Vigny, and J. Massoulié. 1979. Asymmetric and globular forms of acetylcholinesterase in mammals and birds. *Proc. Natl. Acad. Sci.* **76**:2546–2550.

Brimijoin, S., K. Skau, and M.J. Wiermaa. 1978. On the origin and fate of external acetylcholinesterase in peripheral nerve. *J. Physiol.* **285**:143–158.

Brownstein, M.J., J.M. Saavedra, J. Axelrod, G.H. Zeman, and D.O. Carpenter. 1974. Coexistence of several putative neurotransmitters in single identified neurons of *Aplysia*. *Proc. Natl. Acad. Sci.* **71**:4662–4665.

Cammelli, E., A.M. DeBellis, and A. Nistri. 1974. Distribution of acetylcholine and of acetylcholinesterase activity in the nervous tissue of the frog and of the leech. *J. Physiol.* **242**:88–90P.

Coggeshall, R.E. 1972. Autoradiographic and chemical localization of 5-hydroxytryptamine in identified neurons in the leech. *Anat. Record.* **172**:489–498.

Coggeshall, R.E. and D.W. Fawcett. 1964. The fine structure of the central nervous system of the leech, *Hirudo medicinalis*. *J. Neurophysiol.* **27**:229–289.

Coggeshall, R.E., S.A. Dewhurst, D. Weinreich, and R.E. McCaman. 1972. Aromatic acid decarboxylase and choline acetylase activities in a single identified 5-HT containing cell of the leech. *J. Neurobiol.* **3**:259–265.

Cottrell, G.A. 1974. Serotonin and free amino acid analysis of ganglia and isolated neurones of *Aplysia dactylomelia*. *J. Neurochem.* **22**:557–559.

———. 1976. Does the giant cerebral neurone of *Helix* release two transmitters: ACh and serotonin? *J. Physiol.* **259**:44–45P.

Ehinger, B., B. Falck, and H.E. Myhrberg. 1968. Biogenic monoamines in *Hirudo medicinalis*. *Histochemie* **15**:140–149.

Evans, P.D., E.A. Kravitz, B.R. Talamo, and B.G. Wallace. 1976. The association of octopamine with specific neurones along lobster nerve trunks. *J. Physiol.* **262**:51–70.

Falck, B., N.-A. Hillarp, G. Thieme, and A. Torp. 1962. Fluorescence of catecholamines and related compounds condensed with formaldehyde. *J. Histochem. Cytochem.* **10**:348–354.

Fambrough, D.M. 1979. Control of acetylcholine receptors in skeletal muscle. *Physiol. Rev.* **59**:165–227.

Flacke, W. and T.S. Yeoh. 1968. The action of some cholinergic antagonists and anticholinesterase agents on the dorsal muscle of the leech. *Br. J. Pharmacol. Chemother.* **33**:145–153.

Fuchs, P.A., J.G. Nicholls, and D.F. Ready. 1981. Membrane properties and selective connexions of identified leech neurons in culture. *J. Physiol.* **316**:203–223.

Fühner, J. 1917. Ein Vorlesungsversuch zur Demonstration der erregbarkeitssteigernden Wirkung des Physostigmins. *Arch. Exp. Pathol. Pharmokol.* **82**:81–85.

Gaskell, J.F. 1914. The chromaffine system of annelids and the relation of this system to the contractile vascular system in the leech, *Hirudo medicinalis*. *Philos. Trans. R. Soc. Lond. B* **205**:153–212.

Gerschenfeld, H.M. 1973. Chemical transmission in invertebrate central nervous systems and neuromuscular junctions. *Physiol. Rev.* **53**:1–119.

Gerschenfeld, H.M. and D. Paupardin-Tritsch. 1974. Ionic mechanisms and receptor properties underlying the responses of molluscan neurones to 5-hydroxytryptamine. *J. Physiol.* **243**:427–456.

Giller, E. and J.H. Schwartz. 1971a. Choline acetyltransferase in identified neurons of abdominal ganglion of *Aplysia californica*. *J. Neurophysiol.* **34**:93–107.

———. 1971b. Acetylcholinesterase in identified neurons of abdominal ganglion of *Aplysia californica*. *J. Neurophysiol.* **34**:108–115.

Goodman. C.S., M. O'Shea, R. McCaman, and N.C. Spitzer. 1979. Embryonic development of identified neurons: Temporal pattern of morphological and biochemical differentiation. *Science* **204**:1219–1222.

Hall, Z.W. 1973. Multiple forms of acetylcholinesterase and their distribution in endplate and non-endplate regions of rat diaphragm muscle. *J. Neurobiol.* **4**:343–361.

Hall, Z.W., M.D. Bownds, and E.A. Kravitz. 1970. The metabolism of gamma aminobutyric acid in the lobster nervous system. Enzymes in single excitatory and inhibitory axons. *J. Cell Biol.* **46**:290–299.

Hanley, M.R., G.A. Cottrell, P.C. Emson, and F. Fonnum. 1974. Enzymatic synthesis of acetylcholine by a serotonin-containing neurone from *Helix*. *Nature* **251**:631–633.

Henderson, L. 1981. Serotonergic transmission between isolated leech neurones in culture. *Neurosci. Abstr.* (in press)

Hildebrand, J.G., J.G. Townsel, and E.A. Kravitz. 1974. Distribution of acetylcholine, choline, choline acetyltransferase and acetylcholinesterase in regions and single identified axons of the lobster nervous system. *J. Neurochem.* **23**:951–963.

Hildebrand, J.G., D.L. Barker, E. Herbert, and E.A. Kravitz. 1971. Screening for neurotransmitters: A rapid radiochemical procedure. *J. Neurobiol.* **2**:231–246.

Hökfelt, T., O. Johansson, A. Ljungdahl, J.M. Lundberg, and M. Schultzberg. 1980. Peptidergic neurones. *Nature* **284**:515–521.

Ito, T. 1936. Zytologische Untersuchungen ueber die Ganglienzellen des Japanischen medizinischen Blutegels, *Hirudo nipponica*, mit besonderer Berucksichtigung auf die "dunkle Ganglienzelle." *Okajimas Folia Anat. Jpn.* **14**:111–169.

Kehoe, J. 1972. Three acetylcholine receptors in *Aplysia* neurones. *J. Physiol.* **225**:115–146.

Kerkut, G.A., C.B. Sedden, and R.J. Walker. 1966. The effect of DOPA, methyl DOPA and reserpine on the dopamine content of the brain of the snail, *Helix aspersa*. *Comp. Biochem. Physiol.* **18**:921–930.

―――――. 1967a. Cellular localization of monoamines by fluorescence microscopy in *Hirudo medicinalis* and *Lumbricus terrestris*. *Comp. Biochem. Physiol.* **21**:687–690.

―――――. 1967b. A fluorescence microscopic and electrophysiological study of the giant neurones of the ventral nerve cord of *Hirudo medicinalis*. *J. Physiol.* **189**:83–85P.

Kravitz, E.A., B.A. Battelle, P.D. Evans, B.R. Talamo, and B.G. Wallace. 1976. Octopamine neurons in lobsters. *Neurosci. Symp.* **1**:67–81.

Kuffler, D.P. 1978. Neuromuscular transmission in longitudinal muscle of the leech, *Hirudo medicinalis*. *J. Comp. Physiol.* **124**:333–338.

Kuffler, S.W. and J.G. Nicholls. 1966. The physiology of neuroglial cells. *Ergeb. Physiol. Biol. Chem. Exp. Pharmakol.* **57**:1–90.

Kupfermann, I. 1979. Modulatory actions of neurotransmitters. *Annu. Rev. Neurosci.* **2**:447–465.

Kuzmina, L.V. 1968. Distribution of biogenic monoamines in the nervous system of the body segments of the leech, *Hirudo medicinalis*. *J. Evol.*

Biochem. Physiol. (Suppl., Physiology and biochemistry of invertebrates), pp. 50–56.

Le Douarin, N.M. and M.-A.M. Teillet. 1974. Experimental analysis of the migration and differentiation of neuroblasts of the autonomic nervous system and of neuroectodermal mesenchymal derivatives, using biological cell marking technique. *Dev. Biol.* **41**:162–184.

Lent, C.M. 1973. Retzius cells: Neuroeffectors controlling mucus release by the leech. *Science* **179**:693–696.

Lent, C.M., J. Ono, K.T. Keyser, and H. Karten. 1979. Identification of serotonin within vital-stained neurons from leech ganglia. *J. Neurochem.* **32**:1559–1563.

Livingstone, M.S., R.M. Harris-Warrick, and E.A. Kravitz. 1980. Serotonin and octopamine produce opposite postures in lobsters. *Science* **208**:76–79.

MacDonald, J.F. J.L. Barker, D.L. Gruol, L.M. Huang, and T.G. Smith. 1979. Desensitizing excitatory responses to peptides, purines, protons and drugs revealed using cultured mammalian neurons. *Soc. Neurosci. Symp.* **5**:592. (Abstr.).

MacIntosh, F.C. and W.L.M. Perry. 1950. Biological estimation of ACh. In *Methods in medical research* (ed. R. W. Gerard), pp. 78-92. Year Book, Chicago.

Mann, K.H. 1961. *Leeches (Hirudinea)*. Pergamon Press, New York.

Marsden, C.A. and G.A. Kerkut. 1969. Fluorescent microscopy of the 5-HT and catecholamine-containing cells in the central nervous system of the leech *Hirudo medicinalis*. *Comp. Biochem. Physiol.* **31**:851–862.

Mason, A., A.J. Sunderland, and L.D. Leake. 1979. Effects of leech Retzius cells on body wall muscles. *Comp. Biochem. Physiol.* **63C**:359–361.

Massoulié, J., S. Bon, and M. Vigny. 1980. The polymorphism of cholinesterase in vertebrates. *Neurochem. Int.* **2**:161–184.

McAdoo, D.J. and R.E. Coggeshall. 1976. Gas chromatographic-mass spectrometric analysis of biogenic amines in identified neurons and tissues of *Hirudo medicinalis*. *J. Neurochem.* **26**:163–167.

McCaman, M.W., D. Weinreich, and R.E. McCaman. 1973. The determination of picomole levels of 5-hydroxytryptamine and dopamine in *Aplysia, Tritonia* and leech nervous tissue. *Brain Res.* **53**:129–137.

Muller, K.J. and U.J. McMahan. 1976. The shapes of sensory and motor neurones and the distribution of their synapses in ganglia of the leech: A study using intracellular injection of horseradish peroxidase. *Proc. R. Soc. Lond. B* **194**:481–499.

Muller, K.J., S.A. Scott, and B.E. Thomas. 1978. Specific associations between sensory cells. *Carnegie Inst. Washington Yearbook* **77**:69–70.

Osborne, N.N., G. Briel, and V. Neuhoff. 1972. The amine and amino acid composition in the Retzius cells of the leech *Hirudo medicinalis*. *Experentia* **28**:1015–1018.

Perez, H.V.Z. 1942. On the chromaffin cells of the nerve ganglia of *Hirudo medicinalis* Lin. *J. Comp. Neurol.* **76**:367–401.

Poll, H. and A. Sommer. 1903. Ueber phaeochrome Zellen in Centralnervensystem des Blutegels. *Arch. Anat. Physiol. (Physiol. Abt.) 1903* (Nr. X).

Potter, L.T. 1970. Synthesis, storage and release of (^{14}C) acetylcholine in isolated rat diaphragm muscles. *J. Physiol.* **206**:145–166.

Ready, D. and J. Nicholls. 1979. Identified neurones isolated from leech CNS make selective connections in culture. *Nature* **281**:67–69.

Rude, S. 1969. Monoamine-containing neurons in the central nervous system and peripheral nerves of the leech, *Hirudo medicinalis*. *J. Comp. Neurol.* **136**:349–371.

Rude, S., R.E. Coggeshall, and L.S. Van Orden III. 1969. Chemical and ultrastructural identification of 5-hydroxytryptamine in an identified neuron. *J. Cell Biol.* **41**:832–854.

Sargent, P.B. 1975. "Transmitters in the leech central nervous system: Analysis of sensory and motor cells." Ph.D. thesis, Harvard University, Cambridge, Massachusetts.

――――. 1977. Synthesis of acetylcholine by excitatory motoneurons in central nervous system of the leech. *J. Neurophysiol.* **40**:453–460.

Sargent, P.B., K.-W. Yau, and J.G. Nicholls. 1977. Extrasynaptic receptors on cell bodies of neurons in the central nervous system of the leech. *J. Neurophysiol.* **40**:446–452.

Sawada, M. and R.E. Coggeshall. 1976. Ionic mechanism of 5-hydroxytryptamine induced hyperpolarization and inhibitory junctional potential in body wall muscle cells of *Hirudo medicinalis*. *J. Neurobiol.* **7**:63–73.

Sawada, M., J.M. Wilkinson, D.J. McAdoo, and R.E. Coggeshall. 1976. The identification of two inhibitory cells in each segmental ganglion of the leech and studies on the ionic mechanism of the inhibitory junctional potentials produced by these cells. *J. Neurobiol.* **7**:435–445.

Schnaitman, C., V.G. Erwin, and J.W. Greenawalt. 1967. The submitochondrial localization of monoamine oxidase. An enzymatic marker for the outer membrane of rat liver mitochondria. *J. Cell Biol.* **32**:719–735.

Schneider, F.H. and C.N. Gillis. 1965. Catecholamine biosynthesis *in vivo*: An application of thin layer chromatography. *Biochem. Pharmacol.* **14**:623–626.

Stuart, A.E., A.J. Hudspeth, and Z.W. Hall. 1974. Vital staining of specific monoamine-containing cells in the leech central nervous system. *Cell Tissue Res.* **153**:55–61.

Viallia, M. 1934. Le cellule cromaffini dei gangli nervosa negli Irudinei. *Atti. Soc. Ital. Sci. Nat.* **73**:57–73.

Von der Wense, T. 1939. Ueber den nachweis von Adrenalin in Wurmern and Insekten. *Pfluegers Arch.* **241**:284–288.

Walker, R.J., G.N. Woodruff, and G.A. Kerkut. 1968. The effect of acetylcholine and 5-hydroxytryptamine on electrophysiological recordings from muscle fibres of the leech *Hirudo medicinalis*. *Comp. Biochem. Physiol.* **24**:987–990.

――――. 1970. The action of cholinergic antagonists on spontaneous excitatory potentials recorded from the body wall of the leech, *Hirudo medicinalis*. *Comp. Biochem. Physiol.* **32**:690–701.

Wallace, B.G. 1976. The biosynthesis of octopamine—Characterization of lobster tyramine β-hydroxylase. *J. Neurochem.* **26**:761–770.

――――. 1981. Distribution of AChE in cholinergic and non-cholinergic neurons. *Brain Res.* (in press).

Wallace, B.G., B.R. Talamo, P.D. Evans, and E.A. Kravitz. 1974. Octopamine: Selective association with specific neurons in the lobster nervous system. *Brain Res.* **74:**349–355.

Weiss, K.R., J.L. Cohen, and I. Kupfermann. 1978. Modulatory control of buccal musculature by a serotonergic neuron (metacerebral cell) in *Aplysia*. *J. Neurophysiol.* **41:**181–203.

Willard, A.L. 1981. Effects of serotonin on the generation of the motor program for swimming by the medicinal leech. *J. Neurosci.* (in press).

Yaksta-Sauerland, B.A. and R.E. Coggeshall. 1973. Neuromuscular junctions in the leech. *J. Comp. Neurol.* **151:**85–99.

Zipser, B. 1980. Identification of specific leech neurones immunoreactive to enkephalin. *Nature* **283:**857–858.

9
Development of the Nervous System

David A. Weisblat
Department of Molecular Biology
University of California, Berkeley
Berkeley, California 94720

Leeches were a focus of interest of the 19th century pioneers of modern experimental embryology. In the 1880s, Charles O. Whitman, the original director of the Woods Hole Marine Biological Laboratory, presented the first analysis of developmental cell lineage by describing the successive cleavages leading from the fertilized egg to the embryo of the glossiphoniid leech (Whitman 1878, 1887, 1892). On the basis of his observations, Whitman stated the idea that each identified cell of an early embryo, and the clone of its descendant cells, plays a specific, predestined role in later development. Embryological interest in leeches declined at the turn of the century. (Whitman's subsequent work on animal behavior laid the groundwork for the discipline of ethology, while his student T.H. Morgan took up *Drosophila* genetics and revolutionized the science of heredity.) Revival of leech embryology was initiated recently by Roy Sawyer and Juan Fernandez.

Today, because of the detailed knowledge we have of the functional elements of the adult leech nervous system, we can ask specific, focused questions regarding neural development. For instance, do the neurons of a given functional class share a common line of descent in embryogenesis? Are sensory and motor neuron fields established over broad territories at first and only later restricted to the adult territories? Are the synaptic connections between the sensory, interneurons, and

FIGURE 1

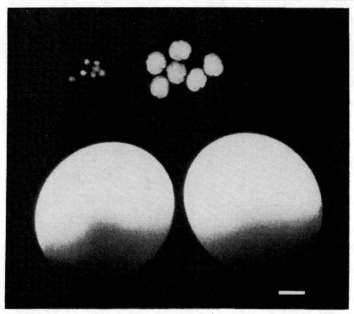

▲ Leech embryos. A size comparison of the embryos of *H. triserialis* (6, upper right), *H. medicinalis* (7, upper left), and *H. ghilianii* (2, bottom). Scale bar, 500 μm.

motor neurons and the muscles responsible for a given behavioral routine formed in a particular temporal order?

To answer such questions, studies of leech neurodevelopment were begun in the mid-1970s. The first task was to find a leech that could be cultivated in the laboratory to provide a constant supply of suitable embryos. Sawyer examined many species and found two favorable candidates. One, *Helobdella triserialis*, is native to California and feeds on the body fluids of snails. It reaches an adult length of about 1 cm and propagates with an egg-to-egg generation time of about 6 weeks. The other, *Haementeria ghilianii*, is native to French Guiana and feeds on the blood of mammals. It reaches an adult length of about 50 cm and has an egg-to-egg generation time of about 10 months (Sawyer et al. 1981). *Helobdella*'s short generation time, simplicity of cultivation, and hardy embryo make it favorable for developmental studies, but its small size renders it less favorable for studying neurophysiology (see Fig. 1). In contrast, the enormous size of *Haementeria* makes even its embryonic nervous system accessible to techniques that require penetrations of single cells; but the long generation time, demanding breeding conditions, and the more fragile embryo present drawbacks when compared with *Helobdella*. Fortunately, both species belong to the same family (Glossiphoniidae); thus, despite differences in size and habit, they are similar in adult body plan and embryonic development. For many embryological purposes, the two species can be considered as interchangeable, providing a greater scope for experimentation than offered by either species alone. Another glossiphoniid species, *Theromyzon rude*, combines the hardiness of *Helobdella* with the large size of *Haementeria* (Fernandez 1980; Fernandez and Stent 1980). Unfortunately, it feeds preferentially within ducks' noses and has not been induced to breed in the laboratory. *Hirudo medicinalis*, the species upon which most neurophysiological work reported in this book has been done, shares many features with the glossiphoniid leeches, even though it is of a different family (Hirudinidae). However, its embryos are too small to be useful for certain types of experiments and its early embryonic development differs from that of the Glossiphoniidae (see below). Accordingly, most of the information in this chapter was obtained using *Helobdella* and *Haementeria*.

Helobdella and *Haementeria* lay eggs that are about 0.5 mm and 2.5 mm in diameter, respectively, and are rich in yolk. The eggs are laid in clutches, enclosed in transparent cocoons, that remain attached to the ventral body wall of the brooding parent. Embryonic development begins as soon as the eggs are laid. Two weeks (*Helobdella*) or a month (*Haementeria*) later, a juvenile leech has arisen whose form differs from the adult mainly by its smaller size. The egg yolk provides the nutrients needed for embryonic development; upon exhaustion of the yolk, typically the juvenile leech, still attached to the underside of its parent, rides to a host animal for its first meal. Postembryonic growth and maturation of the nervous system represent primarily an increase in cell size rather than cell number.

EMBRYOGENESIS

The embryonic development of glossiphoniid leeches has been divided into 11 stages, beginning with egg deposition and extending up to the point at which the juvenile leech is ready for its first meal (Fernandez 1980; Weisblat et al. 1980a; Stent et al. 1982); some stages have been

FIGURE 2

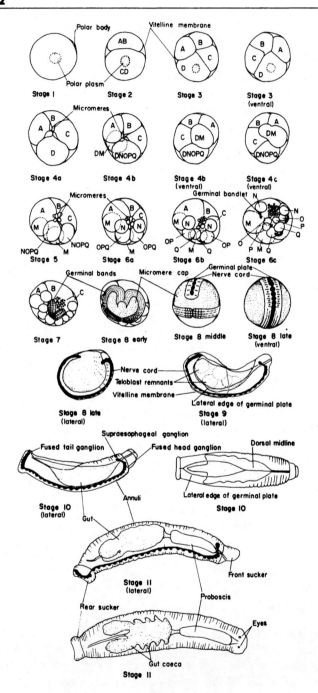

further divided into substages. The definition of the stages and substages is based on morphological criteria discernable in the living embryo. Figure 2 is a schematic representation of the stages of development of *Helobdella*. The criteria for the beginning and end of each stage are given in Table 1. These stages are equally applicable to *Haementeria*. When first laid, the glossiphoniid leech egg is filled throughout with colored yolk and enclosed in a clear vitelline membrane. As the egg approaches its first cleavage (midstage 1), however, a region of yolk-free cytoplasm, or teloplasm, appears at each of its two poles. The poles mark the future dorsal and ventral surfaces of the embryo. The first (meridional) cleavage divides the egg into cells designated AB and CD (stage 2); most or all of the teloplasm passes into cell CD. The second cleavage (also meridional) gives rise to four large cells, macromeres A, B, C, and D (stage 3); D receives most of the teloplasm. The macromeres bud off an unknown number of smaller cells, the micromeres (stage 4a); then cell D cleaves roughly equatorially to yield cells designated DNOPQ and DM (stage 4b). Cell DNOPQ lies dorsally and receives the teloplasm from the dorsal pole; cell DM lies ventrally and receives the ventral teloplasm. At this stage, separation of the embryo into the three germinal layers has been accomplished; the progeny of cells A, B, and C give rise to endoderm, the progeny of DNOPQ give rise to ectoderm, and the progeny of DM give rise to mesoderm (Whitman 1887). The next two cleavages establish the bilateral symmetry of the embryo: DM divides to yield the pair of left and right M cells (stage 4c); DNOPQ divides to yield the pair of NOPQ cells lying on either side of the future midline (stage 5). Three further cleavages of the NOPQ cell pair produce four bilateral cell pairs designated as N, O, P, and Q (stages 6a, 6b, and 6c).

The paired M, N, O, P, and Q cells, among which the teloplasm of the egg is partitioned, are referred to as teloblasts. Each teloblast carries out a series of 30–100 highly unequal divisions, producing a germinal bandlet of small stem cells "like bullets from a machine gun" (J. Fernandez, unpubl.). Most of the teloplasm, but little of the yolk, from each teloblast is eventually distributed to its stem cell progeny. The bandlets produced by the five teloblasts on each side grow and merge on either side of the midline to form a prominent pair of cell ridges, the right and left germinal bands. The mesodermal stem cell

◀ Stages of glossiphoniid embryogenesis schematically illustrated with *H. triserialis*. Unless otherwise indicated, all views are from the dorsal aspect. The criteria for the various stages are listed in Table 1.

TABLE 1

Stages of glossiphoniid embryogenesis

Stage	Duration
Stage 1 — uncleaved egg	from the time of egg laying to the onset of first cleavage
Stage 2 — two cells	from the onset of first cleavage to the onset of second cleavage
Stage 3 — four cells	from the onset of second cleavage to the onset of micromere cleavage
Stage 4a — micromere quartet	from the onset of micromere cleavage to the onset of D macromere cleavage to form cells DM and DNOPQ
Stage 4b — macromere quintet	from the onset of D macromere cleavage to the onset of cleavage of cell DM
Stage 4c — mesoteloblast formation	from the onset of cleavage of cell DM to the onset of cleavage of cell DNOPQ
Stage 5 — ectoteloblast precursor	from the onset of cleavage of cell DNOPQ to the onset of cleavage of the NOPQ cell pair.
Stage 6a — N teloblast formation	from the onset of cleavage of cell NOPQ to the onset of cleavage of cell OPQ to form cells OP and Q
Stage 6b — Q teloblast formation	from the onset of cleavage of cell OPQ to the onset of cleavage of cell OP to form teloblasts O and P
Stage 6c — O and P teloblast formation	from the onset to the completion of the cleavage of cell OP to form teloblasts O and P
Stage 7 — germinal band formation	from the completion of cleavage of the OP cell pair to the onset of coalescence of left and right germinal bands
Stage 8 — germinal band coalescence	from the onset to the termination of coalescence of left and right germinal bands
Stage 9 — nerve cord coalescence	from the termination of germinal band coalescence to the coalescence of the 32nd pair of segmental hemiganglionic primordia
Stage 10 — body closure	from the formation of the 32nd segmental ganglion to the completion of fusion of the lateral edges of the germinal plate along the dorsal midline
Stage 11 — yolk exhaustion	from the completion of body closure to the exhaustion of the yolk in the embryonic gut

A schematic representation of each stage is shown in Fig. 2.

bandlet produced by M lies under the four ectodermal bandlets. With ongoing stem cell production, right and left germinal bands advance along mirror-image, crescent-shaped paths over the dorsal surface of the embryo (stage 7) and converge at the site of the future head.

Meanwhile, as the germinal bands continue to lengthen, their midportions move apart circumferentially over the dorsal surface. Eventually, right and left germinal bands enter the ventral surface; finally they meet and coalesce, like a zipper, from the future head rearward on the ventral midline (stage 8). The coalescing germinal bands form the germinal plate, which lies along the ventral midline. Before migration of the germinal bands, the superficial ectodermal bandlets lie in the order q, p, o, n from medial to lateral. (The germinal bandlets and the stem cells they comprise are designated by the appropriate lowercase letter; capital letter designations are reserved for the teloblasts themselves.) The circumferential migration of the germinal bands reverses this order so that n is medial and q is lateral in the coalescing bands. Thus, the left and right n bandlets come into apposition at the ventral midline within the germinal plate; this led Whitman (1878) to suggest that the nerve cord is derived from the descendants of the n stem cells.

This description of embryogenesis up to the stage of germinal plate formation (end of stage 8) seems valid for glossiphoniid leeches in general, but not for the Hirudinidae, such as *Hirudo*. *Hirudo* eggs are much smaller (less than 100 μm in diameter). They contain little yolk and are deposited in a hard-shelled cocoon that contains an albuminous fluid. The embryo forms a larval mouth and muscles with which it swallows the surrounding fluid. It increases in size, and the cells are large enough for physiological experiments soon after germinal plate formation. As with glossiphoniid leeches, embryogenesis in *Hirudo* proceeds via five pairs of teloblasts that produce stem cells, germinal bands, and a germinal plate. But in the hirudinid embryo, germinal bands form on the ventral (rather than the dorsal) surface and zipper together directly, without undergoing the circumferential migration seen in the Glossiphoniidae (Schleip 1936). Subsequent development of the germinal plate is much the same in *Hirudo* as in the glossiphoniids (J. Fernandez, unpubl.).

The cells of the germinal plate proliferate to form adult tissues. This cell proliferation results in a gradual thickening and expanding of the germinal plate over the surface of the embryo back into dorsal territory. Early in the expansion process (stage 9), the germinal plate becomes partitioned along its length into a series of tissue blocks. Each block corresponds to a future body segment. Segmentation starts at the front and progresses rearward, and by the time the hindmost

FIGURE 3

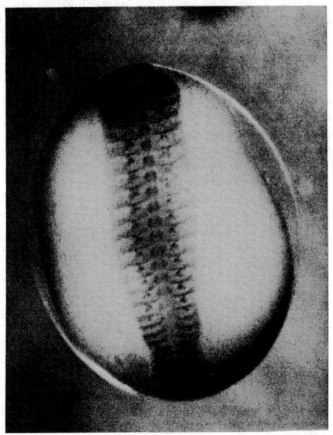

▲ Haematoxylin-stained, stage-9 *Haementeria* embryo. Expansion of the germinal plate has covered about one-fourth of the ventral surface. The segments are separated by transverse septa. In the front third of the germinal plate, each segment contains an intact ganglion on the ventral midline. The paired dark spots lying on either side of the ganglion are nephridia. In the middle third of the plate, the segments contain ganglionic primordia at various stages of coalescence, and in the rear of the plate ganglionic primordia are not yet manifest. (Photo by R.T. Sawyer.)

segment has formed (end of stage 9), the expanding germinal plate covers about one-third of the ventral surface. When the expanding germinal plate covers about half the ventral surface, the embryo hatches from the vitelline membrane. Eventually (stage 10), right and left leading edges of the expanding germinal plate meet and coalesce on the dorsal midline, closing the leech body. Meanwhile, formation of the gut is underway. It first appears as a cylinder (filled with yolk provided by macromeres A, B, and C and the remnants of the teloblasts) and then becomes segmented by a series of annular constrictions. The constrictions give rise to paired gut lobes, or caeca, in register with the abdominal segments. Gut segmentation is completed at body closure (end of stage 10); the embryo now has the general shape of the adult leech. When the yolk in the gut is exhausted (end of stage 11), the juvenile leech is ready for its first meal.

NEUROGENESIS

The genesis of the segmental nervous system follows the general developmental sequence of the germinal plate, proceeding from front to rear (J. Fernandez, pers. comm.; A. Kramer and D. Stuart, pers. comm.). First, hemispheric masses of cells appear on both sides of the germinal plate (ventral) midline in each segment. Each pair of massed cells constitutes the primordium of a segmental ganglion. The paired masses merge one after the other on the ventral midline to form globular ganglia, which already contain about the same number of neurons as adult ganglia (see Fig. 3). When the expanding germinal plate covers about half the ventral surface of the embryo, all 32 ganglia of the ventral nerve cord are present. At first, the ganglion lies in direct contact with its anterior and posterior neighbors. Later, the abdominal ganglia become separated by short connective nerves, which elongate as the distance between ganglia increases during growth. The anterior four and posterior seven ganglia fail to separate, however, and fuse to form the head (subesophageal) ganglion and the tail ganglion, respectively. The supraesophageal ganglion has an embryonic origin distinct from that of the segmental ganglion (see below).

NEURONAL DIFFERENTIATION

A.P. Kramer (pers. comm.) has shown that it is possible to remove the nervous system from *Haementeria* embryos, penetrate its neurons with micropipettes, take intracellular recordings, and inject markers such as Lucifer Yellow. Using these techniques, it has been found that when the segmental ganglia first appear in their globular form, the neurons have not yet grown axons and dendrites; groups of these immature neurons are coupled via junctions that permit injected dye to pass from cell to cell. Later, the neurons begin to grow axons; their dye-coupling largely disappears, but they still show no action potentials. It is difficult to ascertain the exact time of initiation of electrical activity because it is always possible that the small resting potentials and inexcitability of the fragile embryonic neurons result from the trauma of impalement. It is clear, however, that action potentials appear no later than when growing axons have begun to enter the connective and segmental nerves. Finally, by the time of body closure the embryonic nervous system has taken on its general adult properties. Neurons at this stage exhibit apparent excitatory and inhibitory postsynaptic potentials, indicating that synaptic connections have formed (A. Kramer and J. Kuwada, unpubl.).

Electron microscopic examination of the nervous systems of *Helobdella* and *Haementeria* at various stages of development indicates: (1) that glial ensheathment of nerve cell bodies and processes occurs after formation of interganglionic connectives and (2) that synapses in the leech central nervous system (CNS) do not acquire their adult morphology (see Chapter 6) until well after movements have begun (J. Fernandez, pers. comm.; K. Muller and D. Weisblat, unpubl.).

The origin of neuronal branching patterns has been examined by Lucifer Yellow injection of identified neurons of *Hirudo* embryos in late embryogenesis (B. Wallace, unpubl.). Motor neurons (annulus erector [AE]) and sensory neurons (touch [T]) initially send out more processes than are maintained by their adult homologs (see Fig. 4); later, selective loss of some processes occurs. Thus, at least for these neurons, segmental variation in the adult branching pattern arises from differential process loss rather than differential outgrowth during development.

NEUROTRANSMITTER SYNTHESIS

By the time of body closure, the embryonic neurons have begun to synthesize transmitters. This has been shown by analysis of transmitter synthesis and accumulation in *Haementeria* embryos by using radiochemical techniques to detect acetylcholine (ACh), serotonin (5-hydroxytryptamine [5-HT]), and γ-aminobutyric acid (GABA) and fluorescence methods to detect 5-HT and dopamine (see Chapter 8) (H. Cline and D. Stuart, unpubl.).

The capacity for ACh synthesis is still very low when neurons within the assembled ganglia have begun to grow axons. The synthetic capacity rises rapidly, however, once the neurons have sent axons into the connective and segmental nerves and have begun to produce action potentials. By the time of body closure (end of stage 10), the capacity of the nerve cord for ACh synthesis has increased more than 25-fold. The capacity for synthesis of other neurotransmitters studied follows approximately the same time course; it is not known whether there is a temporal hierarchy in the initiation of synthesis of different transmitters.

BEHAVIOR

The maturation of the leech nervous system is paralleled by an evolution of the embryo's behavior, which proceeds by a stereotyped sequence of motor acts (A. Kramer, unpubl.). Videotape recordings of *Haementeria* embryos have shown that embryonic behavior progresses from simple, irregular twitching to complex movement patterns. Simple early movements form part of more complex later movements. These more complex movements are, in turn, components of locomotory routines of the juvenile and adult animals. Some of these embryonic movements may fulfill a physiological role, such as circulation of nutrients or hatching from the egg membrane, but others might be just incidental to the formation of functional connections within the developing nervous system. Overt movements begin with peristaltic contractions as germinal plate formation nears completion (late stage 8). Since at this point there is no nervous system, the peristalsis is likely to be due to myogenic contraction of cells girding the embryo; but whether these contractile cells are embryonic transients or forerunners of adult circular muscles is unknown. Peristalsis eventually results in the embryo's hatching from the egg membrane. After hatching, the peristalsis soon ceases and the embryo begins intermittent lateral bending movements, using left and right longitudinal muscles. By this time (stage 10), maturation of the ventral nerve cord has progressed sufficiently to permit execution of neurogenic movements (i.e., muscle contractions elicited by the nervous system). After body closure (stage 11), the lateral bending gives way to a more complicated cycle of movements that is evidently the prelude to the "inchworm walking" of the juvenile leech (see Fig. 5). After front and rear suckers have developed, this ripens into actual walking. By the time it is a juvenile (end of stage 11), the leech can also swim (see Chapter 7).

FIGURE 4

A

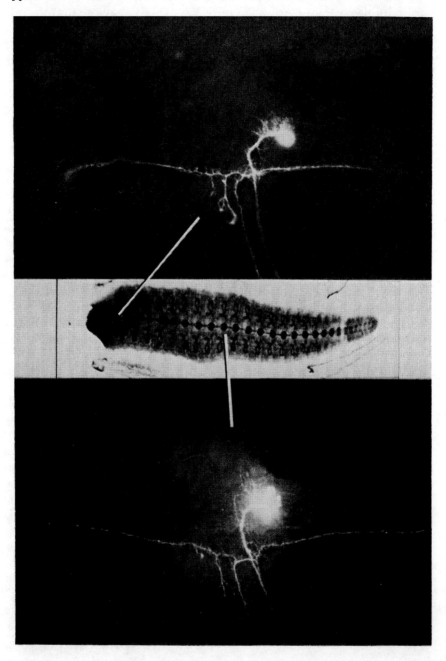

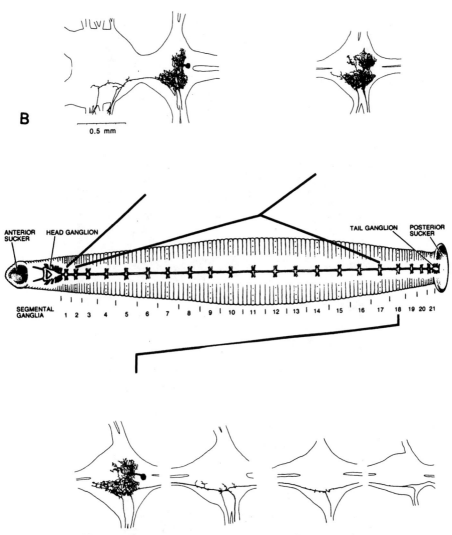

◀▲ Reduction of the arborization of the AE motor neuron during development. (A) Lucifer Yellow-filled AE cells in one anterior and one midbody ganglion of an 8-day-old *Hirudo* embryo. Note that both cells send processes into the anterior and posterior connectives, as well as out the contralateral segmental nerves. (B) The branching pattern of AE cells in anterior, midbody, and posterior ganglia of an adult *Hirudo*. Evidently, anterior AE cells lose their rear-going branch, posterior AE cells their front-going branch, and midbody AE cells both their connective branches at some point in development. Inserts in both panels show the ganglia of origin of the filled cells. Anterior is to the left. (Figure courtesy of Bruce Wallace.)

FIGURE 5

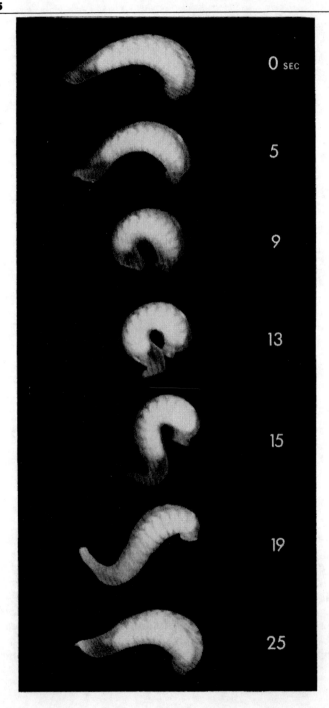

CELL LINEAGE

Whitman and his students described the lines of descent from the uncleaved egg to the teloblasts and their stem cell bandlets by direct microscopic examination. But with the increase in cell number later in development, the fate of individual cells cannot be followed by direct observation. More detailed cell pedigrees in the developing leech nervous system can be obtained using a new technique (Weisblat et al. 1978, 1980a,b). A tracer molecule is injected into an identified cell of the early embryo. Embryonic development is allowed to progress to a later stage, at which time the distribution pattern of the tracer within the tissues is observed. This method is useful if three conditions are met: (1) embryonic development continues normally after injection of the tracer; (2) the injected tracer remains intact and is not diluted too much in the developing embryo; and (3) the tracer does not pass through junctions linking embryonic cells and is passed on exclusively to descendants of the injected cell.

One such tracer is the red-fluorescing dye rhodamine, which is coupled to a dodecapeptide carrier. The carrier peptide is synthesized from amino acids in the unnatural D-configuration to render it resistant to digestion by proteolytic enzymes. The rhodamine peptide (RDP) is particularly useful for visualizing the development of the germinal plate (S. Zackson, unpubl.). For this purpose, a teloblast or teloblast precursor cell is injected with RDP; the embryo is later fixed in the presence of the blue-fluorescing, DNA-specific stain Hoechst 33258 (to distinguish individual cells) and cleared. Under the fluorescence microscope, blue-fluorescing nuclei lying within the confines of red-fluorescing cells belong to progeny of the injected cell; the orientation of the spindle axis of cells caught in mitosis can be determined by inspection. The time at which a stem cell cleaves (as judged by the number of younger stem cells separating it from the parent teloblast)

◄ Embryonic locomotion. Prewalking movement of early stage-11 *Haementeria* embryo recorded on videotape. The step cycle consists of shortening, downward bending, upward bending, and body elongation. Later in stage 11, upon maturation of front and rear suckers, this movement ripens into actual walking. For walking, at the start of the step cycle the elongated leech is attached to the solid substrate by its rear sucker. The front sucker then becomes attached and the rear sucker released; now shortening and downward bending follows, after which the rear sucker becomes reattached at a new, more advanced position and the front sucker is released. The step is completed by upward bending and body elongation. (Figure courtesy of A.P. Kramer.)

and the orientation of the spindle axis for at least the first few stem cell divisions are stereotyped for each teloblast line.

Using this technique, it has also been seen that the mesodermal m bandlets subdivide into clusters of labeled cells early in germinal band formation. Such subdivision is the first overt sign of segmentation of the mesodermal tissue and occurs before the coalescence of right and left germinal bands. Within adjacent clusters, the positions and sizes of the cells are the same. This suggests that the early development of the mesodermal body segment proceeds, as does cleavage and teloblast formation in the early embryo, by a sequence of stereotyped cell divisions. Each mesodermal cell cluster is derived from a single m stem cell and is probably the precursor of a single mesodermal segment on that side.

Another tracer that meets the criteria for establishing cell lineages is the enzyme horseradish peroxidase (HRP). HRP-stained embryos can be examined intact or, in greater detail, after embedding in plastic and sectioning. In a typical experiment, an identified teloblast is injected with HRP before, or soon after, stem cell production begins. The injected embryos are fixed and stained for HRP just before body closure (stage 10) 5–7 days later; at this point, the ganglia are well-formed, but neurons have not yet undergone sufficient differentiation to allow their identification on the basis of size and position alone. It has not been possible to examine the HRP distribution in completely differentiated neurons because the intensity of the HRP stain declines in older embryos.

Figure 6 presents four embryos, in each of which a different ectodermal precursor teloblast on the right side had been injected during stage 7. Evidently, the HRP-stained descendants of each injected teloblast form a distinct pattern that is repeated from segment to segment (cf. Fig. 3). (This pattern is the same in every embryo in which a given teloblast has been injected.) The stain pattern is confined to the same side as the injected teloblast and does not cross the midline. (Injection of the left teloblasts results in stain patterns that are in mirror image of those shown in Fig. 6.) Thus, there is little or no lateral migration of cell bodies across the ventral midline. Examination of sections of embryos such as those shown in Figure 6 shows that each teloblast contributes some cells to the ventral nerve cord. In the case of the N teloblast, the progeny are almost exclusively within the ganglia of the ventral nerve cord. Descendants of O, P, and Q teloblasts, however, give rise to characteristic patterns of cell clusters both in the ventral nerve cord and in the lateral ectoderm; the sizes and positions of these clusters are sufficiently invariant from segment to segment and from specimen to specimen so that the identity of the injected teloblast is

FIGURE 6

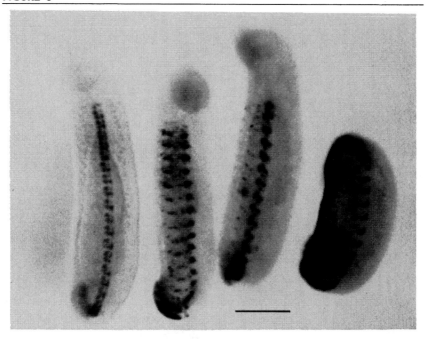

▲ Distribution of cells derived from individual teloblasts. Ventral view of four embryos in which the right-hand N, P, O, and Q teloblasts, respectively, were injected with HRP early in stage 7. The embryos were fixed and stained 6 days later (stage 10). The Q injected embryo appears shorter because it was bent when fixed. The stain pattern is confined to the same side of the germinal plate as the injected teloblast in each embryo. Except in the Q injected embryo, the pattern has a sharp anterior boundary between the unstained tissue derived from the last unlabeled stem cell and the stained tissue derived from the first labeled stem cell. Anterior is up. Scale bar, 200 μm.

apparent from the stain pattern in the embryo (D.A. Weisblat and S.-Y. Kim, unpubl.). With few exceptions, all of the cells of the segmental ganglion (including glia) are derived from the N, O, P, and Q teloblasts. The few exceptional cells seem to be derived from the M teloblast and may be nonneuronal. These experiments also showed that the cells of the supraesophageal ganglion (see Chapter 4) do not arise from the teloblasts at all. Even when the ectoteloblast precursor NOPQ is injected to insure that all ectodermal stem cells are HRP-labeled, the supraesophageal ganglion fails to stain. Further experi-

FIGURE 7

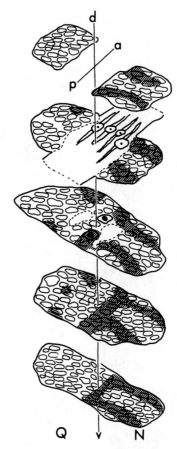

▲ Embryonic origins of the cells of the *Helobdella* segmental ganglion. The drawing shows five horizontal sections through a schematized midbody segmental ganglion of the embryo at the stage at which the germinal plate covers nearly all of the ventral surface. (Dorsal aspect at the top; front edge facing away.) The two pairs of dark, elongated contours in the center of the second section from the top represent identifiable muscle cells in the longitudinal nerve tract; they are descendants of the M teloblasts. The two dark, circular contours in the center of the middle section represent two identifiable glial cells, each of which is a descendant of one N teloblast. The light contours do not correspond to actual cells but are shown to indicate the approximate size, disposition, and number of neurons in the ganglion. In the left side of the ganglion hatching indicates domains containing descendants of the left Q teloblast; in the right half of the ganglion, hatched domains contain descendants of the right N teloblast; domains not hatched on either side contain descendants of the O and P teloblasts.

ments revealed that its neurons arise from the macromeres, presumably via the micromeres produced in stage 4a (Weisblat et al. 1980a).

In the embryos shown in Figure 6, the teloblasts had been injected with HRP after their stem cell production had begun. In these embryos, there is a sharp anterior boundary of the HRP stain pattern between the tissue (unstained) arising from the last stem cell produced before the HRP injection and the tissue (stained) arising from the first stem cell produced after the injection. The sharpness of this boundary suggests that there is little or no longitudinal migration of cell bodies within the germinal plate. The variation in position of the stain boundary relative to the ganglion borders allows an indirect determination of the number of stem cells from a given teloblast that give rise to a single hemiganglion. If a single stem cell founded one hemiganglion, the stain boundary seen in embryos injected after the initiation of stem cell production should always coincide with a ganglion border. If, however, more than one stem cell from a given teloblast contributes to a single hemiganglion, then sometimes a hemiganglion at the stain boundary will receive progeny from both labeled and unlabeled stem cells. If this happens, and since (as shown above) cells do not migrate longitudinally, the stain boundary should sometimes fall within a ganglion. The greater the number of stem cells contributing to a single ganglion, the greater should be the proportion of intrasegmental stain boundaries. When such experiments were carried out with the N teloblast, it was found that the stain boundary fell in only two positions, either between ganglia or midway through a ganglion (Weisblat et al. 1980a). Therefore, it can be concluded that more than one, and probably only two, n stem cells contribute progeny to a single hemiganglion.

The major ganglionic subpopulations arising from the ectoteloblasts are distributed as follows: from N, two transverse slabs of cells in the anterior and posterior regions of the hemiganglion and a longitudinal band of cells near the midline of the ventral part of the ganglion (Fig. 7); from O, an oblique column of cells extending from the midportion of the ventral aspect through to the anterior portion of the dorsal aspect of the ganglion; from P, a thin, transverse band of cells near the center of the ventral aspect of the ganglion; and from Q, two small patches of cells near the midline at the anterior and posterior edges of the ganglion (Fig. 7). The N and O teloblasts each contribute many more cells to the ganglion than do P and Q.

In view of the stereotyped location of the progeny of each teloblast in the embryonic ganglia and the constant positions of identified neurons in the ganglia of the adult leech (see Chapter 4), it seems likely that each identified neuron normally arises from a given telo-

FIGURE 8

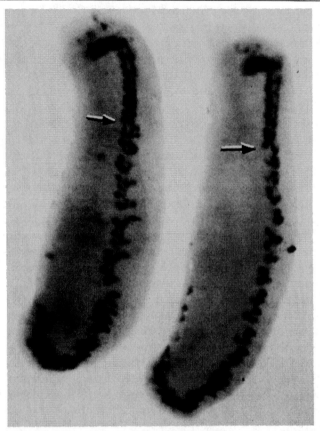

▲ Teloblast ablation. Ventral view of two *Helobdella* embryos whose right N teloblasts were injected with HRP at stage 6a (soon after they formed) and whose left N teloblasts were ablated by DNase injection early in stage 7. The embryos were fixed and stained 5 days later (stage 10). (→) The bilateral stain distribution within the nerve cord below this point indicates that in the absence of the progeny of one N teloblast, progeny of the surviving N teloblast are found on both sides of the ganglia. Anterior is up. The embryos are rolled slightly to the side to show the subesophageal ganglion angling to the left at the beginning of the nerve cord in each embryo.

blast. The observation that the neuropil glia are derived from the N teloblasts supports this hypothesis (Weisblat et al. 1980a). But the small size of embryonic *Helobdella* neurons and the practice of fixing the tissue prior to visualizing with HRP, makes it impossible to ascertain the teloblast of origin of electrophysiologically identified neurons (such as T, pressure [P], or nociceptive [N]). For that purpose, RDP is being used for in vivo experiments on the much larger *Haementeria* embryos. RDP is injected into a teloblast of a *Haementeria* embryo in stage 6 or 7. When the embryo has developed further (stage 9 or 10), the red-fluorescing, RDP-labeled neurons are penetrated with micropipettes for electrophysiological and anatomical identification. In this fashion, it should prove possible to learn the teloblast of origin of identified neurons in the leech CNS. The segmental pattern of progeny from the various teloblasts is homologous between *Haementeria* and *Helobdella* (D.A. Weisblat, unpubl.).

ABLATION

The cell lineage analyses show that leech neurogenesis is highly determinate, in the sense that a particular teloblast gives rise to a particular part of the nervous system during normal development. But what about abnormal development? Upon death of any given cell, can its normal developmental role be taken over by any other cell? It has been shown that killing a teloblast of the early embryo leads to characteristic developmental aberrations (Weisblat et al. 1980b; Blair and Weisblat 1982; Blair 1981). For instance, killing an N teloblast by injection of protease (see Appendix A) or deoxyribonuclease (DNase) results in an embryonic nervous system whose segmental ganglia show anatomical deficits on one side. However, the pattern of these deficits is not as regular as one might expect; the hemiganglia on the ablated side vary in size within the same specimen, ranging from normal to total absence. Hemiganglia that seem normal contain the cells normally provided by the O, P, and Q teloblasts on the same side and, surprisingly, cells derived from the (nonablated) N teloblast on the other side (see Fig. 8); these latter cells have crossed the midline during formation of the ganglionic primordia. That some hemiganglia are entirely absent in these specimens, i.e., do not even receive cells from the (nonablated) O, P, and Q teloblasts on the same side, suggests that the abnormal midline crossing of N-derived cells fluctuates and that the N-derived cells play some organizing role for the O-, P-, and Q-derived cells in founding the ganglionic primordia.

As expected, killing an M teloblast results in an embryo that lacks all mesodermal segmental structures on the ablated side (Blair 1981). However, in addition, the ablated side also lacks recognizable hemiganglia. Here the N, O, P, and Q teloblasts still produce their bandlets of stem cells, but in the absence of the segmental mesodermal tissue blocks, the precursor cells of the nervous system are not organized into ganglionic primordia. These deficits produced by ablation of teloblasts show the importance of interactions between cells of different lines in shaping the development of the leech nervous system.

CONCLUSION

Only incomplete answers of a mainly descriptive nature are currently available to the grand question: "How do neurons of the leech nervous system and their specific interconnections arise during development?" From these answers, however, focused questions pertaining to mechanisms can be formulated. For instance, the ventral part of the egg's teloplasm is given to mesodermal precursor cell DM and the dorsal part to ectodermal precursor cell NDOPQ. Are there differences between the different parts of the teloplasm that determine the different fates of these cells and their progeny? How do the cells descended from the ectodermal precursors N, O, P, and Q and lying within each segment assemble to form the neurons of the ganglionic primordia? How do structural, physiological, and chemical properties of the developing neurons depend on their interaction with other cells or tissues? Does the gradual perfection of the walking and swimming movements require practice, or do motor circuits governing locomotory behavior arise autonomously during embryogenesis without functional feedback? And, does a neuron's line of descent govern its ultimate character because of a particular set of intracellular determinants distributed in successive embryonic cell divisions, or because the cell occupies a particular position in the embryo thanks to the regular cleavage pattern through which it arose? Some of these questions should be answerable using experimental techniques already at hand.

REFERENCES

Blair, S.S. 1981. Interactions between mesoderm and ectoderm in segment formation in the embryo of a glossiphoniid leech. *Dev. Biol.* (in press).

Blair, S.S. and D.A. Weisblat. 1982. Ectodermal interaction during neurogenesis in the glossiphoniid leech, *Helobdella triserialis*. *Dev. Biol.* (in press).

Fernandez, J. 1980. Embryonic development of the glossiphoniid leech *Theromyzon rude*: Characterization of developmental stages. *Dev. Biol.* 76:245-262.

Fernandez, J. and G.S. Stent. 1980. Embryonic development of the glossiphoniid leech *Theromyzon rude*: Structure and development of the germinal bands. *Dev. Biol.* 78:407-434.

Sawyer, R.T., F. LePont, D.K. Stuart, and A.P. Kramer. 1981. Growth and reproduction of the giant glossiphoniid leech *Haementeria ghilianii*. *Biol. Bull.* **160:**322–331.

Schleip, W. 1936. Ontogenie der Hirudineen. *In Klassen und Ordnungen des Tierreichs* (ed. H.G. Bronn), vol. 4, Div. III, Book 4, Part 2, pp. 1-121. Akademie Verlagsgesellschaft, Leipzig.

Stent, G.S., D.A. Weisblat, S.S. Blair, and S.L. Zackson. 1982. Cell lineage in the development of the leech nervous system. In *Neuronal development* (ed. N. Spitzer). Plenum, New York. (In press.)

Weisblat, D.A., R.T. Sawyer, and G.S. Stent. 1978. Cell lineage analysis by intracellular injection of a tracer enzyme. *Science* **202:**1295–1298.

Weisblat, D.A., G. Harper, G.S. Stent, and R.T. Sawyer. 1980a. Embryonic cell lineages in the nervous system of the glossiphoniid leech *Helobdella triserialis*. *Dev. Biol.* **76:**58–78.

Weisblat, D.A., S.L. Zackson, S.S. Blair, and J.D. Young. 1980b. Cell lineage analysis by intracellular injection of fluorescent tracers. *Science* **209:**1538–1541.

Whitman, C.O. 1878. The embryology of *Clepsine*. *Q. J. Micros. Sci.* **18:**215–315.

―――――. 1887. A contribution to the history of germ layers in *Clepsine*. *J. Morphol.* **1:**105–182.

―――――. 1892. The metamerism of *Clepsine*. In *Festschrift zum 70, Geburtstage R. Leuckarts*, pp. 385–395. Engelmann, Leipzig.

10

Regeneration and Plasticity

Kenneth J. Muller
Department of Embryology
Carnegie Institution of Washington
Baltimore, Maryland 21210

John G. Nicholls
Department of Neurobiology
Stanford University School of Medicine
Stanford, California 94305

A striking example of the ability of the central nervous system (CNS) to repair itself after injury is provided by experiments on leeches in which one of the connectives has been transected. About 4 weeks later, when the wound has healed, it is often almost impossible to detect under the dissecting microscope where the original cut had been made (Fig. 1). Moreover, cells that had been connected before transection become reconnected with a high degree of precision (see Muller 1979). Similar experiments also reveal an inherent ability of the leech CNS to modify the pattern and efficacy of its connections. For some time after one connective has been transected and the other has been left intact, animals do not swim or walk normally — head and tail fail to move in a coordinated rhythm. But 4 weeks after transection, the movements are again normal whether or not regeneration of the severed connectives has occurred. Even in animals in which the two cut ends of one of the two connectives remain separated, well-coordinated swimming movements can be observed, as though connections within the CNS somehow have changed, adapting in response to the unrepaired lesion. Comparable recovery phenomena following lesions have been observed in other invertebrates, including crickets, crustaceans, and snails. In the leech, it has been possible to carry the analysis of such repair processes to the cellular level; more-

FIGURE 1

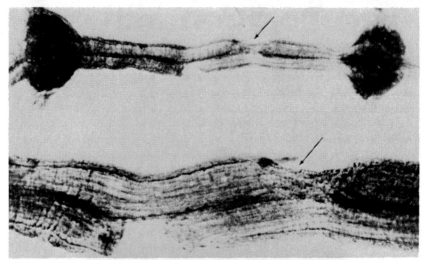

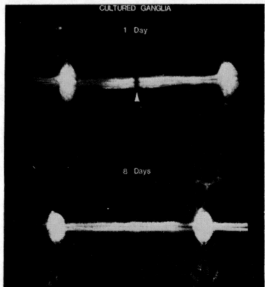

▲ (*Above*) Photographs of a preparation in which one connective in the animal had been cut (↓) 40 days earlier. The connective is largely healed. The lower connective was cut just before taking the photograph to show an acute lesion. (The diameter of a connective is approximately 100 μm.) (*Below*) Dark-field photographs of a cultured preparation 1 day and 8 days after crushing the connectives (▲). Note the extent of the repair at the site of the lesion. (Reprinted, with permission, from Baylor and Nicholls 1971 and Wallace et al. 1977.)

over, after deletion of a single neuron in its entirety, another, homologous cell can take over its role (see Appendix A).

The advantages of the leech for such regeneration studies accrue not only from the stereotyped simplicity of the ganglia and their neuronal connections, but also from the ability of isolated ganglia to survive and regenerate these connections in culture. In a variety of preparations, including leeches with severed connectives, single cells deleted from the CNS by protease injection, and isolated ganglia, or even in isolated single identified nerve cells maintained in culture for several weeks, it has been possible to answer some significant questions about how individual cells grow back toward their targets and re-form connections after their axons have been severed by a lesion. For instance, do the growing axons head unerringly for the "correct" neurons, or do they meander? Do neurons make only correct connections, or do they also make erroneous ones that are later retracted? Does fusion of a cut axon occur with its surviving distal stump, as is the case in crustaceans? And, most importantly, what is the mechanism by which one neuron recognizes another so that it is able to form synapses selectively with that specific target in preference to others?

For long-term changes, the range of problems explored in the leech has been more limited than in crickets, where habituation, conditioning, and sensory deprivation have been studied in detail (Kandel 1976). Apart from the study of such phenomena as short-term synaptic facilitation and depression and presynaptic modulation of transmitter release (Chapter 6), the principal emphasis of the work on leeches thus far has been on long-term changes following lesions in the CNS.

HOW ACCURATELY CAN NEURONS BECOME RECONNECTED DURING REGENERATION?

Electrophysiological recordings provide a convenient test for whether regenerating neurons have reestablished connections with their original targets (Baylor and Nicholls 1971; Jansen and Nicholls 1973; Wallace et al. 1977; Muller 1979). Using such tests, it has been shown that after the connectives have been cut or crushed, touch (T), pressure (P), and nociceptive (N) cells in one ganglion reconnect with their original target sensory and motor cells in the next ganglion. Thus, about 4 weeks after the operation, impulses in a regenerated T cell evoke synaptic potentials in the T cells of the next ganglion. Similarly, impulses in regenerated N and P cells evoke monosynaptic potentials in the L motor neuron of the next ganglion (Fig. 2). Moreover, isolated ganglia removed from the leech can be maintained in a suitable culture

FIGURE 2

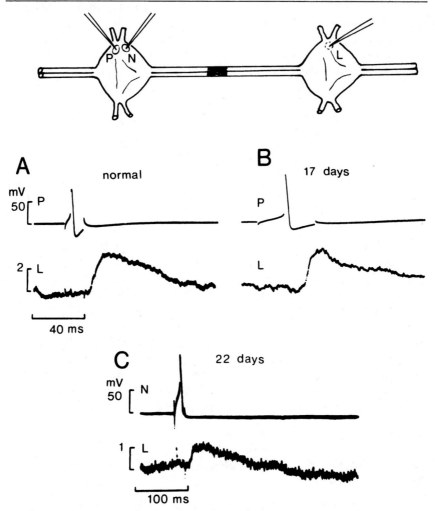

▲ Regenerated connections between individual cells in adjacent ganglia in culture. (A) In normal preparations, an impulse in a P or N (not shown) sensory cell evokes an excitatory potential in the ipsilateral L cell of the next ganglion. (B) After regeneration (17 days in culture), the excitatory potential in the L cell resembles in its latency and amplitude that seen normally, suggesting that the connection may be monosynaptic. (C) In another preparation maintained in culture for 22 days after crushing connectives, an N cell (shown by HRP in Fig. 3 to have been regenerated) evoked synaptic potentials in the L cell, as in a normal preparation. (Reprinted, with permission, from Wallace et al. 1977.)

medium for several weeks, and after connective transection, the cells will regenerate in a manner indistinguishable from that seen in situ (Wallace et al. 1977). It should be noted, however, that following transection the balance of excitation and inhibition in the regenerated connectives may be characteristically shifted. For instance, after regeneration, inhibitory postsynaptic potentials (IPSPs) are evoked in the longitudinal (L) motor neuron by stimulating P or N cells in the next posterior ganglion (Baylor's Law). These IPSPs, which are barely detectable before the operation, are superimposed on and can mask the expression of the normal direct excitatory postsynaptic potentials (EPSPs) and are mediated by an as yet unidentified axon that has also grown across the lesion (Jansen et al. 1974). Moreover, after the operation, synaptic potentials can also be recorded occasionally in cells not ordinarily excited or inhibited by P or N cells.

Do such "abnormal" synaptic potentials represent errors of regeneration? Electrophysiological techniques are inadequate for examining the precision of regenerated connections in detail, since in practice it is not possible to impale all cells of a ganglion to determine how many of them give abnormal responses to impulses in T, P, or N cells of the adjacent ganglion. Even if such an experiment were possible, it would be difficult to interpret its results in terms of precision of regeneration, or lack thereof. For example, if only in a regenerated preparation, but not in the normal animal, stimulation of an N cell in one ganglion gives rise to a synaptic potential in another cell, this need not represent an error in regeneration; the same connection might be present anatomically in unoperated animals but be too weak to be discerned by recording electrically. Alternatively, these connections might not exist at all in unoperated animals but could have arisen as a result of cutting or crushing the connectives. As will be shown, such abnormal connections can arise from axonal sprouting that occurs at a distance from the lesion, even in the absence of regeneration.

GROWTH TOWARD THE TARGET

Anatomical observations have made it possible to determine exactly how a regenerating cell grows to contact its previous targets, how in the meantime those targets have changed, and how new and old synapses compare in structure and distribution.

Fifty years ago, Ramon y Cajal (1928) described a now classic sequence of regeneration in silver-stained preparations of the nervous system of higher vertebrates after crushing a peripheral nerve. He

FIGURE 3

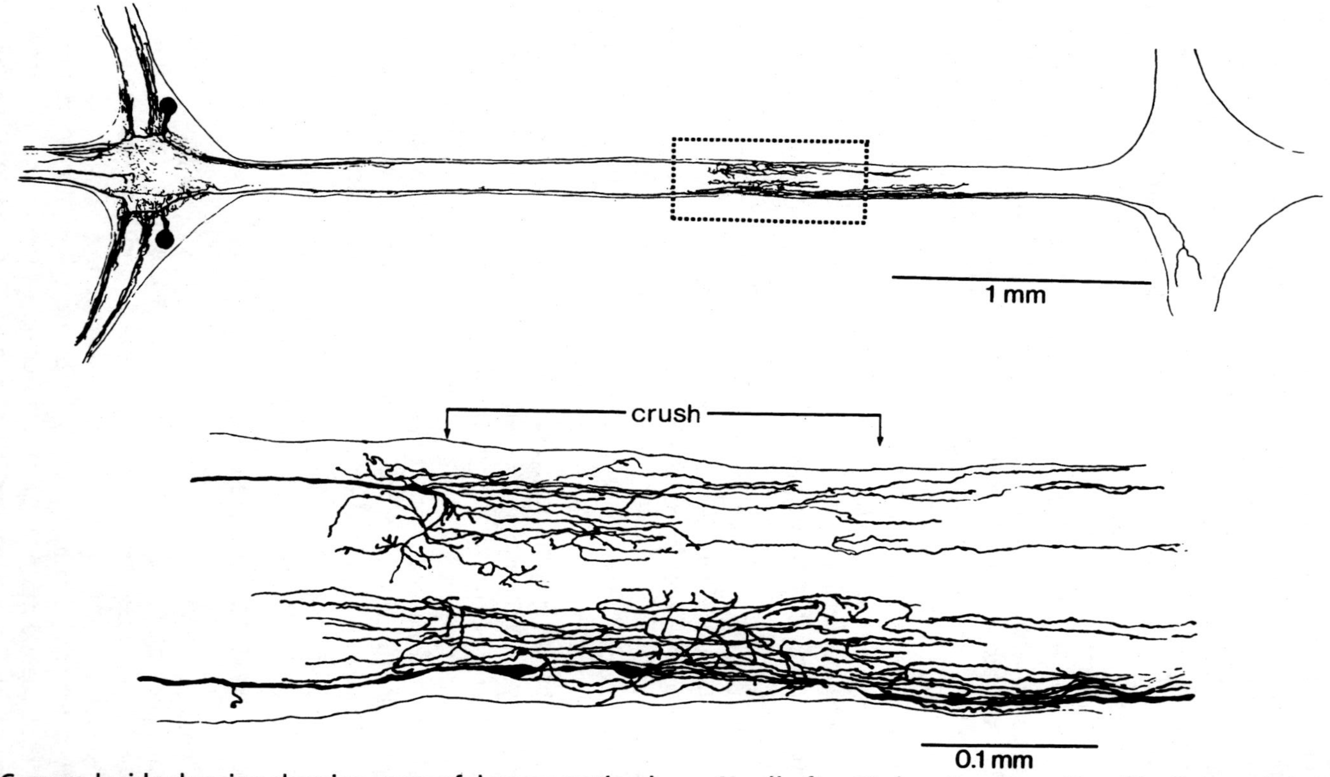

▲ Camera lucida drawing showing successful regeneration by an N cell after 22 days in culture. Two N cells were injected with HRP. Both ramified extensively at the site of the crush and sent branches towards the next ganglion, one of which regenerated into the neuropil, where it arborized. Recordings from this cell are shown in Fig. 2. (Reprinted, with permission, from Wallace et al. 1977.)

observed that regenerating neurons proximal to the crush sprouted abundantly, with fine neurites tipped with growth cones coursing in all directions from the severed ends. Meanwhile, the proximal stumps themselves shrank in caliber. Many of the sprouting branches crossed the crush, a seemingly disorganized region that also supported growth of Schwann cell macrophages. Those nerve fibers that reached across the crush to the distal stump grew along what was presumably the original Schwann cell conduit to their normal target. Once they arrived at the target, the regenerated axons increased in caliber, the diverse sprouting branches near the crush began to disappear, and the nerve resumed its normal appearance.

Since Cajal's time, electron microscopy has provided a more detailed description of some structures involved in regeneration, such as the reticulum within the growth cone, but it is with the study of identified neurons in the leech that our picture of the regeneration process and the accuracy with which it occurs has been carried a step further (Wallace et al. 1977; Muller and Carbonetto 1979; Muller 1979). Within a day or two after connectives in the leech are cut or crushed, sprouts are visible at the proximal ends of the severed axons, in a manner similar to that described by Cajal. Sprouts tipped with growth cones emerge from the severed axon and grow in many different directions; some of them even grow perpendicular to the axon or backwards toward the ganglion in which the axon originated. Although sprouting within the lesion seems to lack directionality, those fibers that reach across the lesion enter the connective occupied by the distal stump, and one or more processes grow, while forming branches, toward the next ganglion (Fig. 3). The amount of branching within the connective seems to vary with cell type and is correlated with the normal extent of branching of that cell type within the ganglion. Thus, within the lesioned connective, N cells as a rule branch more profusely than P cells, which in turn branch more profusely than T cells. Similar growth of processes occurs in organ culture. Within a few weeks after the operation, when physiologically detectable synaptic connections have been reestablished with neurons in the adjacent ganglion, the regenerated neuron has extended processes into it. Intact sensory neurons send only a single axon down the connective toward the next ganglion, so that when several branches of a cell grow along different regions of the connective, not all can be following the original route.

Within days of crushing the connective, synapses form at the site of the crush (Fig. 4), an unusual region for synapses to occur in unoperated leeches (Fernandez and Fernandez 1974). Although the synapses persist within the connective for months, it has not yet been possible to record from them electrically. Under the light microscope, one can

FIGURE 4

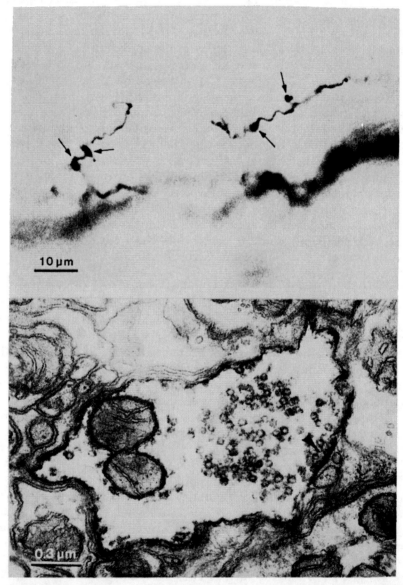

▲ Synapses of T cells are formed de novo at a lesion in the connective. (*Top*) Sprouts of the crushed axon have varicosities (↓) resembling varicosities of the same cell within the ganglion (11 months after crushing). (*Bottom*) Such varicosities are seen by electron microscopy to contain synaptic vesicles and make typical synapses (↙) (3 months after crushing). Reprinted, with permission, from Muller 1979.)

see that stained neurons have occasional enlargements, or varicosities, that are distinct from the growth cones. Synapses are found at some, if not all, of these varicosities. The abnormal synapses made in the connective resemble those made by the same cell within the ganglion, but the varicosities are simpler. Perhaps some swellings that show no synaptic cleft- or membrane-associated densities but contain vesicles are, in fact, nascent synapses. It is also not clear whether neurons that synapse within the connective are synapsing with normal targets. Some motor neurons, such as the L motor neuron, will sprout into the connective following a crush, so that sensory-motor contacts could form at the pseudoneuropil near the site of the lesion.

Although in normal ganglia synapses are confined to the neuropil, synaptogenesis by regenerating cells seems less restricted. Not only do regenerating cells form pseudoneuropils within the connective, but occasionally they also directly contact other cell bodies within the ganglion. It may be the case that when stimulated to grow, neurons form synapses at sites where there are transmitter receptors, as have been shown to exist on denuded cell bodies (Sargent et al. 1977), and that such receptors may be found along axons in the connective.

Examination of regenerating sensory cell axons indicates that some neurons grow directly toward the next ganglion and branch relatively little, if at all; whereas other axons ramify extensively within the pseudoneuropil in the connective, sometimes failing to reach the next ganglion even after a year. It may be the case that once an axon has formed enough synapses within the connective, farther growth toward the next ganglion ceases. In contrast with Cajal's picture of regeneration, some neurons that have successfully connected with targets in the next ganglion retain some sprouts within the connective even after a year, perhaps because of synaptic connections made by those sprouts.

No detailed analysis yet exists of the route and final structure of regenerated axons and their chemical synapses in the leech, but to a first approximation, the structure and distribution of regenerated synapses are identical with those in the normal animal (Fig. 5) (Muller 1979). A regenerated T cell, for example, forms and receives synapses within the next ganglion just as in normal preparations, although there is some indication that the varicose secondary processes are unusually short. A possible explanation is that as a result of denervation, there is extensive sprouting within the target ganglion itself (see also Miyazaki and Nicholls 1976; Miyazaki et al. 1976). Perhaps either the target neurons sprouted toward the incoming sensory neuron, or, alternatively, sensory neurons resident within the ganglion may have sprouted to replace the original synapses.

FIGURE 5

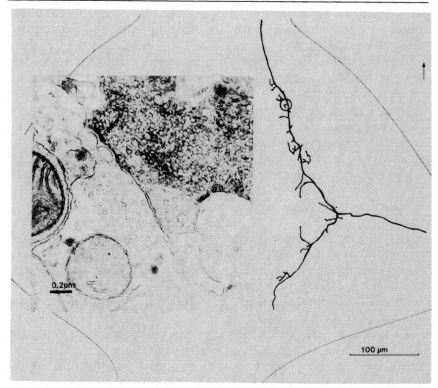

▲ T-cell processes regenerate into posterior adjacent ganglion. Secondary processes resembling those that normally make synapses in posterior ganglion emerge from main axon in this camera lucida tracing of serial 8-μm sections. Preparation was made a year after lesion and represents the end point of regeneration. Anterior at top. (*Inset*) Electron micrograph of secondary process circled in drawing. Varicosity is filled with synaptic vesicles and is presynaptic at arrow. Regenerated process was also postsynaptic (not shown). (Reprinted, with permission, from Muller 1979.)

REGENERATION OF AN ELECTRICAL SYNAPSE: THE S CELL

Neighboring S cells synapse with each other in a single, accessible region of the connective, and the axons of either or both can be recognized without selective staining and may be severed (Chapter 6). S cells thus provide an unusually suitable system for examining (1) the steps taken by a neuron as it regenerates a synaptic contact and (2) the roles of the target, the supporting cells, and the old axon in the formation of that synapse. One would like to know whether, for example, after the axon of one S cell is crushed, the uninjured "target" S cell sprouts in search of a new synaptic contact with the missing but regenerating S cell, or does the target simply await the arrival of the regenerating axon. The answers have come by examining horseradish peroxidase (HRP)-injected S cells in regenerating preparations and simultaneously recording from pairs of cells with intracellular microelectrodes by passing current into one cell and monitoring any coupling potential in the next.

The stages in regeneration of injured S cells exactly parallel those described above by Cajal and seen for sensory neurons (Muller and Carbonetto 1979). Within a few days of crushing or cutting the connective, the S-cell axon sprouts several branches at the crush, seemingly without direction, but only branches that contact the region of the connective that is near the severed distal stump of the S cell proceed down the connective (Fig. 6). Meanwhile, the distal stump survives functionally and structurally intact; several weeks after it has been severed from the S-cell soma, it continues to conduct impulses and excite and be excited by the uninjured target S cell, which does not itself sprout. Complete regeneration of the injured S cell along the distal stump and synapse formation with the adjacent target S cell ordinarily takes 4 or 5 weeks, but the distal stump often forms a pathway that "short circuits" the slower regenerative growth (Carbonetto and Muller 1977). Thus, in about half the examples of regenerating S cells that have been examined in the month after the lesion is made, when S cells have not made direct contact as measured by HRP injection into the regenerate and its target, the regenerating S cell wraps a basket of processes around its own distal stump and synapses with it. At the earliest stages of regeneration during the second week, currents can pass across this junction, but they are apparently insufficient to excite either cell. In the next stage of coupling, impulses in the large distal stump can excite the regenerating S cell, but the fine regenerating processes are unable to excite the distal stump (Fig. 7). Direct passage of current shows that the electrical junctions are nonrectifying, yet impulses conduct in only one direction, presumably because of differences in axon caliber, a phenomenon that has some precedent in the leech (Van Essen 1973; Yau 1976) and elsewhere (Parnas et al. 1969, 1976). When functional regeneration occurs by means of synapses with the distal stump, by about 3 weeks impulses are conducted as they are normally from one cell to the next. Regeneration evidently proceeds toward the target whether or not a synapse is formed with the distal stump, because examination of preparations at later times during the first month reveals that the growing tip of the

FIGURE 6

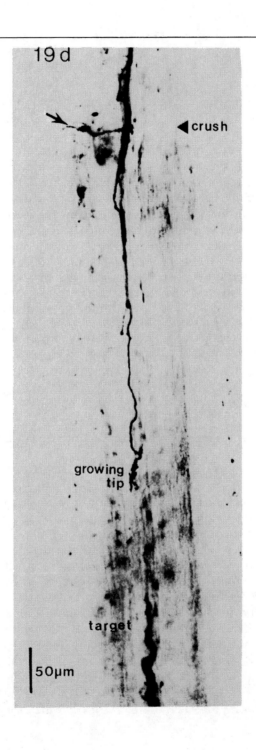

regenerating neuron, filled with characteristic membranous reticulum when viewed in the electron microscope, grows along its own distal stump and closes the gap between regenerating and target-injected processes. Once the target S cell is contacted, synapses are formed and growth stops. Microelectrode recordings reveal that transmission across the new synapse also occurs in stages: (1) coupling with no impulse conduction; (2) one-way conduction of impulses from the target into the fine regenerating neuron but not vice versa across the nonrectifying junction; and (3) as the synapse is strengthened and the caliber of the regenerating neuron increases, impulses propagate both ways across the S-cell synapse as in the final stage of conduction through the distal stump (Fig. 7).

The formation of the synapse between S cells seems to be a trigger for the distal stump to degenerate, for the stump rapidly disappears during the month after regeneration; its fleeting degenerating profile filled with multivesicular bodies and other degeneration remnants is rarely seen. There is good evidence that in the absence of regeneration, the severed distal stump persists for months, slowly shrinking and staining more densely as the glial cell that surrounds it hypertrophies. In addition, since regeneration readily occurs across two crushes that would totally isolate a segment of distal axon, no light has yet been thrown on any single factor that would account for the longevity of the severed stump. Nevertheless, the existence of the distal stump, or some trace of it, is apparently essential for precise regeneration.

WHAT ROLE DOES THE TARGET PLAY IN REGENERATION?

Although the initial directed outgrowth of axons may be independent of the synaptic target, contact with appropriate target cells seems to determine when and where regenerating axons stop growing. If the normal target cells are missing during regeneration, the axons may continue growing until they synapse with suitable but inappropriate targets or the neurons may die. In the CNS, it is not known how

◄ Nineteen days after crushing, an S cell has regenerated along its distal stump to within a few hundred micrometers of the uninjured target S cell, which has also been injected but has not grown. In this micrograph from whole mount, one of the branches at the crush is visible (→). The S cells were not electrically coupled. (Reprinted, with permission, from Muller and Carbonetto 1979.)

FIGURE 7

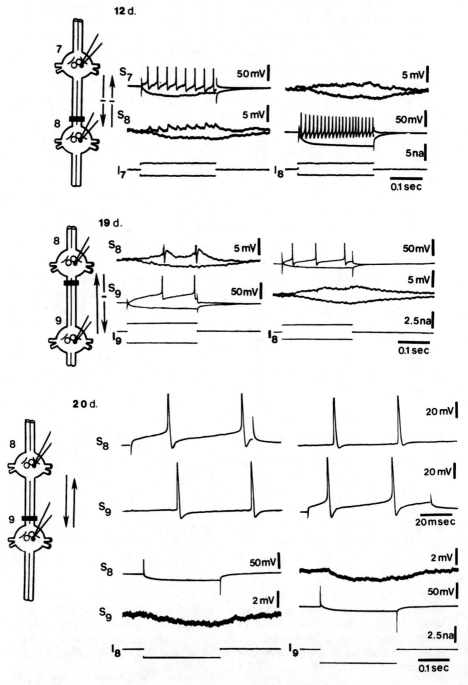

210

mature, uninjured axons would respond to the simple loss of their targets. Most studies that point to target-cell influences in the mammalian CNS have dealt with large populations of presynaptic neurons that innervate large populations of target cells; the elimination or denervation of the targets usually requires disruptive surgical manipulations that sever many axons, including those to be studied. Under these circumstances, it is difficult to determine precisely which neurons regenerate or sprout, what degree of specificity of reinnervation is shown by individual neurons, and which aspect of the surgical intervention prompts the observed response. To investigate how a given axon responds to the loss of its synaptic target, it would therefore be useful to remove selectively the target of a single regenerating (or intact) axon without damaging the axon or disturbing its surrounding environment.

The system of S interneurons in the leech is well-suited for such an investigation, since each S cell axon normally makes an electrical synapse exclusively with the tip of the adjacent S-cell axon, its synaptic target, and can regenerate this connection specifically. When the target of a severed S-cell axon is selectively eliminated by the intracellular injection of protease, the severed axon regenerates along its distal stump in a normal fashion, stopping at the usual site of synapse without making detectable alternative connections (Muller and Scott 1979; Scott and Muller 1980).

◄ Stages of coupling through S-cell distal stump during S-cell regeneration. Electrodes were placed in adjacent S cells; each electrode was used to pass current and measure voltage. Connectives were crushed 12–20 days previously at sites indicated by solid horizontal bar in diagrams at left. All three pairs shown were coupled, and the vertical arrow indicates the direction, if any, in which impulses were conducted from one cell to the next. Within 2 weeks after crushing, the earliest coupling may be recorded (12 days, *top*), but coupling is too weak to sustain impulse transmission. Impulses in distal stump, first generated in S_7, are seen electrotonically as sharp sawteeth in S_8. Depolarizing and hyperpolarizing currents injected into S_7 (I_7) or S_8 (I_8) pass pass equally well into the other cell. The junction is therefore nonrectifying. At the next degree of coupling (19 days, *middle*), impulses can propagate into the regenerating neuron, S_8 (tops of impulses not shown), but the regenerating neuron cannot excite the target S cell (at right). Before the regenerating S cell has grown to its target S cell, impulses may propagate in both directions through the distal stump (20 days, *bottom*). Hyperpolarizing currents spread well in both directions, indicating good coupling through the uninjected stump. (Reprinted, with permission, from Muller and Carbonetto 1979.)

FIGURE 8

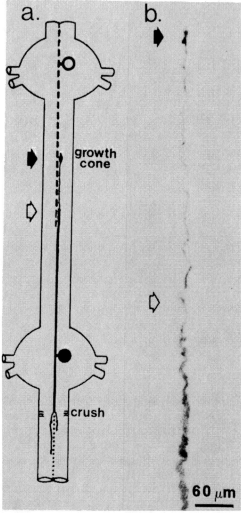

▲ (a) The intact axon of an injured S cell (in the posterior ganglion, lower) sprouts when its target S cell (------------) in the upper ganglion is selectively destroyed. The target S cell was injected with protease, and a distant axon of the adjacent S cell was severed by crushing the connective (crush). The schematic diagram shows that the injured axon regenerated in a normal fashion across the crush and along its severed distal stump (............). The intact axon of that cell, which had been in contact with the killed cell in the old synaptic region (◊), sprouted a fine process tipped with a growth cone (◆). (b) Higher-power view of a sprouted axon 25 days after operation. (Reprinted, with permission, from Scott and Muller 1980.)

Intact S cells are apparently unaffected by the killing of their target with protease; they neither grow nor retract and form no aberrant synapses. In contrast, when its target is killed, an intact axon can be triggered to sprout at its tip by injuring another branch of the same cell (Muller and Scott 1980; Fig. 8). The sprouted axons extend along the pathway formerly occupied by the killed cell, but they make no detectable synapses. For S cells, it appears that cellular injury activates general growth processes within the neuron that are expressed only by axonal branches that have lost contact with their target. The results with sprouting of sensory cells in the periphery described in Chapter 5 show certain differences. There, denervation of skin, produced by killing T or N cells with protease, leads to expansion of the receptive field of the appropriate undamaged T or N cell into the vacant area.

ROLE OF GLIAL CELLS IN REGENERATION

In many systems there is evidence that glial cells can influence the regeneration of axons. For example, in vertebrates, peripheral and central axons will regenerate across a Schwann cell-ensheathed region but not across one ensheathed by central glial cells. In *Hirudo*, the size and stereotyped locations of the glial cells within the CNS make them directly accessible to experimental manipulation during nerve regeneration. For example, a single glial cell ensheathes each of the paired lateral connectives that link adjacent ganglia along the segmented nerve cord; each of these glial cells extends the length of the connective (~5 mm), with the glial nucleus located midway along the connective (see Chapter 4). Crushing or cutting a connective severs not only the axons but also the connective glial cell and divides it into a nucleated portion and an anucleate glial stump.

In operated animals, the fate of the stump and the response of the nucleated portion have been followed by intracellular recording, by intracellular injection of Lucifer Yellow dye and HRP as tracers, and by electron microscopy (Elliott and Muller 1981). The nucleated portion of the glial cell does not divide, degenerate, or grow appreciably. As for the anucleate stump, in Figure 9 it is shown that it does not become reconnected to the nucleated portion, but it does maintain its resting potential and normal morphology for as long as 3 months, after which it begins to deteriorate. Some of the axons degenerate, while the remainder appear morphologically and physiologically normal. Thus, both nucleated and anucleate glial portions persist throughout the 1 to 2 months required for axons to regenerate functional connections. However, after destruction of the connective glial cells by protease

FIGURE 9

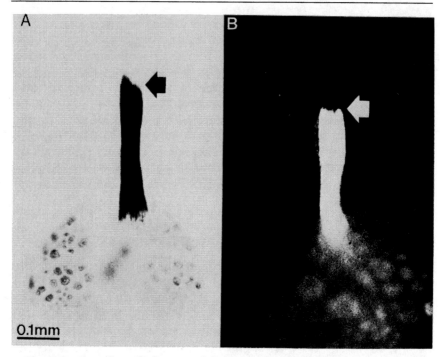

▲ Photographs of a glial stump injected with both HRP (*A*) and Lucifer Yellow dye (*B*). (◆, ◇) Site of a crush 6 weeks previously. No electrical connection had formed with the nucleated portion of the cell lying above the arrow, and thus the dye could not move across the crush as it did into the glial cells within the ganglion. The fluorescent photograph (*B*) was taken of living tissue. The preparation was then fixed and stained for HRP (*A*). Anterior at top. (Reprinted, with permission, from Elliott and Muller 1981).

injection, axons can still regenerate to form appropriate connections in the next ganglion and in the connective (E. Elliott and K. Muller, unpubl.); therefore, the connective glial cell is not necessary for accurate regeneration of synapses.

CHRONIC CHANGES IN SYNAPTIC POTENTIALS

In addition to the long-term changes exhibited by sensory and motor cells following deletion of single cells (described in Chapter 5),

FIGURE 10

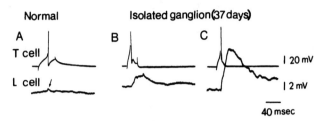

▲ Abnormal chemical synaptic potentials in the L motor neuron following stimulation of a T cell within a ganglion isolated in the animal. (A) The normal response of the L motor neuron, which is a small, electrically mediated potential (↓). (B) The results obtained in a ganglion isolated within the animal 37 days previously by cutting roots and connectives; an impulse in the T cell produced a large, delayed excitatory potential characteristic of a chemical synapse. (C) Hyperpolarizing currents injected into the L cell increased the amplitude of the synaptic potential, confirming that the newly appearing component was initiated chemically. In this record, the L cell had been hyperpolarized by about 30 mV. The current-voltage relation of the L cell is linear, and hyperpolarization has little or no effect on the electrically mediated potential produced by the T cell in normal ganglia (Nicholls and Purves 1970). (Reprinted, with permission, from Jansen et al. 1974.)

changes also occur to otherwise stable synaptic connections between cells within the CNS in the absence of regeneration. Days or weeks after connectives on either side of a ganglion have been cut, synaptic potentials at first appear normal and then become enhanced or, as with some regenerated interganglionic connections, the balance of excitation and inhibition may change (Jansen et al. 1974).

For example, within a ganglion isolated by cutting connectives, the electrical synaptic potential in L motor neurons normally evoked by the T cell acquires an extra, late chemical component (Fig. 10). Cutting the connectives causes the P cell to inhibit strongly the annulus erector (AE) motor neuron rather than to excite it (Fig. 11). An unidentified intervening neuron that is normally only weakly excited by the P cell and inhibits the AE motor neuron evidently becomes excited by the P cell as a consequence of the operation. Accordingly, conditions that interrupt polysynaptic pathways, such as bathing in saline containing high concentrations of Ca^{++} and Mg^{++}, can restore the unusual connections to near normal. Some other connections of the same sensory neuron do not change. Certain synaptic activities, such as the EPSP evoked by the N cell in the L motor neuron, are greatly enhanced and

FIGURE 11

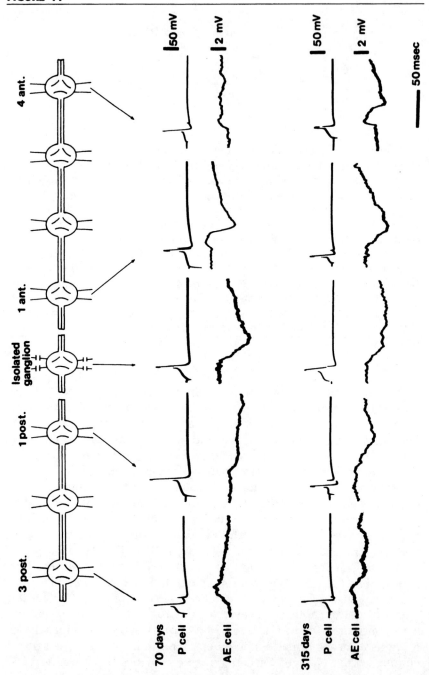

prolonged without measurable change in the input impedance of the motor neuron. It is not known whether strengthening of these synapses is attributable to increased transmitter release at individual terminals, to postsynaptic hypersensitivity to transmitter resulting from a loss of other inputs, or to sprouting of the presynaptic neuron to fill postsynaptic sites unveiled by the lesion.

The plastic changes that occur in the connections between ganglia when connectives are cut develop most rapidly in ganglia near the lesions and, for certain connections, spread sequentially along the length of the animal. Based on their short delay for transmission and their lability in the presence of elevated Ca^{++} and Mg^{++} concentrations, the new synaptic potentials seem to be mediated by rapidly conducting interneurons. It may prove difficult to find these interneurons by intracellular recording from ganglia that exhibit changed connections, because isolated lengths of axon in the leech survive functionally for long periods without their cell bodies (Thompson and Stent 1976; Van Essen and Jansen 1977; Wallace et al. 1977) and are themselves capable of forming new synapses (Calabrese 1977; Carbonetto and Muller 1977).

Because the leech can recover function without regeneration of normal pathways, one would like to know what sort of reorganization within a ganglion replaces the normal circuits. For example, by cutting the right connective anterior to the ganglion and the left connective posterior to it, the fibers through the ganglion are disrupted and the leech is unable to swim or walk until appropriate neuronal growth occurs. Similarly, ganglia completely isolated within the animal are normally incapable of generating a swimming rhythm, but a few

---

◄ With progressively longer times after the connectives have been severed, synaptic potentials in the AE motor neuron became changed in ganglia along the length of the animal. Inhibitory synaptic potentials appeared first in the isolated ganglion after about 1 month and after about 70 days in the neighboring ganglia (1 post. and 1 ant.). The inhibitory synaptic potentials were often preceded by small excitatory potentials. In more distant ganglia, 1 or 2 months after the operation impulses in P sensory cells still produced normal excitatory synaptic potentials (3 post. and 4 ant., 70 days). Later (315 days), inhibitory synaptic potentials in the AE motor neuron appeared in progressively more distant ganglia. Even after a year, however, synaptic potentials in AE cells of ganglia far from the lesion were less changed than those within the isolated ganglion. A similar profile and time course of changes were seen in the effects of N sensory cells upon AE cells. (Reprinted, with permission, from Jansen et al. 1974.)

weeks after disconnection from the other ganglia in the animal, circuits are rearranged so that motor neurons can rhythmically fire as they do to produce swimming (Kristan and Guthrie 1977). For swimming to return, we now know that the reorganization that must occur within the ganglion could, in theory, be relatively small (Chapter 7) but would require specific synapse formation and might occur as an adaptive change in response to injury. The exact nature of the altered connections leading to swimming is not known.

One behavioral change that has been correlated with altered synaptic potentials is puzzling: This is the inhibitory connection that develops from the P cells onto the AE motor neuron. Pressing or pinching the skin normally causes the annuli to erect ridges, but in animals whose connectives have been cut, the same stimulus causes rapid flattening of any erect annuli (Fig. 11). There is some promise that the physiological basis of these synaptic changes will be found, because many of them come about in cultured ganglia with a time course that parallels that of regeneration in the whole animal. The environment can be controlled in culture, indicating that the changes are not brought about by systematic changes from either the stress of altered circulation or lack of rostrocaudal coordination.

SELECTIVE CONNECTIONS OF ISOLATED IDENTIFIED NEURONS IN CULTURE

A considerable simplification for studying the mechanism and specificity of synapse formation, as well as plasticity, can be achieved by working with isolated neurons in culture, where even greater control over the cells and their environment is possible. Preparations used for such studies include explants of brain containing heterogeneous populations of neurons and glia, neurons dissected from the CNS, or peripheral ganglia and muscle fibers maintained in culture. Individual identified nerve cells have been dissected from the CNS of the leech and maintained in culture (Ready and Nicholls 1979; Fuchs et al. 1981). These neurons, which can be isolated from ganglia by a simple mechanical procedure without exposing them to enzymes, offer distinct advantages for studying the selectivity of synapse formation, since a variety of types of nerve cells can be kept alive in culture singly, in pairs, or in groups and then compared with their counterparts in the ganglion.

In such preparations, one can compare the cell membrane properties and fine structure of the isolated cells with those in the animal and examine the synaptic interactions that develop when cells are cultured together. Will cells form synapses in culture at random or selectively? If connections are formed selectively, how will their pattern compare with that of connections seen in the animal? To approach these problems, single identified cells can be placed next to one another to determine the efficacy with which connections are formed and to explore the range of compatible synaptic partners.

Isolated leech neurons in culture also provide simplified preparations for studying the formation of synapses and mechanisms of synaptic transmission. Within the CNS of the leech, as in other invertebrates, experiments on these problems are complicated by the fact that synapses occur in the neuropil at a distance from the cell body (Chapter 6). Hence, intracellular recordings made from the soma do not faithfully mirror the synaptic potentials. For the same reason, currents injected into the presynaptic cell body to modulate transmitter release or to spread through electronic junctions between coupled cells are attenuated and distorted at the terminals. In addition, the synapses are inaccessible for direct exploration by techniques such as iontophoretic application of drugs or transmitters, and sites of contact between cells are buried in a tangle of processes, complicating anatomical studies. In culture, cells can be placed in direct contact, and it is therefore possible to develop synaptic connections in which one can more fully analyze the anatomy, physiology, and pharmacology of regenerated synapses between neurons.

A single cell is removed from the nervous system by the series of steps illustrated in Fig. 12. After opening the connective tissue capsule surrounding the ganglion and washing the glia, a loop of fine nylon thread is slipped over the cell's soma and tied tightly around the initial process. The cell is then pulled out of the ganglion. After 10 minutes, the neuron can be freed by gentle shaking or by snapping the thread, and then transferred by means of a fine suction pipette with a constriction to a dish containing culture medium (Leibowitz 15 with 2% fetal calf serum). Leech neurons that have been torn from their normal surroundings and placed in a foreign environment survive well for several weeks.

During the first days, despite the absence of glia or normal blood, the isolated neurons not only recover, but sprout and form electrical and chemical connections (Fig. 13). In addition, the fine structure of the isolated cells continues to appear normal, and even after several weeks in culture, there is still a striking similarity between the electrophysiological properties of isolated T, P, and N mechanosensory cells and Retzius (R) cells and those of their counterparts in situ (Fig. 14). Hence, the cultured cells preserve their characteristic sets of membrane conductances and maintain a roughly normal distribution of

FIGURE 12

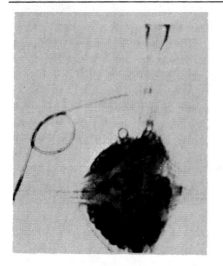
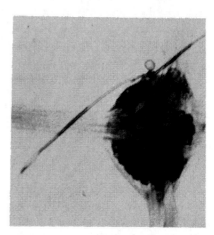
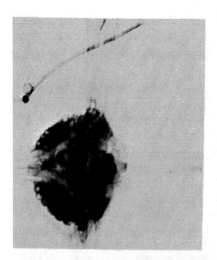

▲ Steps in the removal of an individual N cell from a ganglion using nylon thread of 13 μm diameter. (Reprinted, with permission, from Ready and Nicholls 1979.)

FIGURE 13

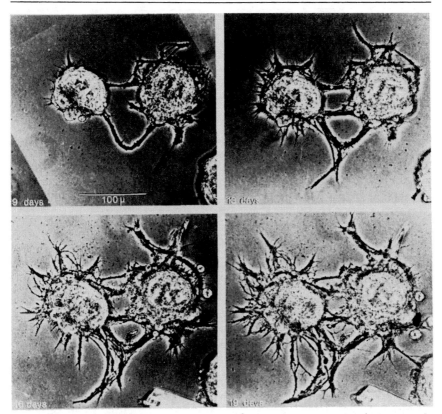

▲ Development of connections in culture between a pair of R cells over 19 days. Note the preferential distribution of processes running between the two cells and the relative paucity of processes running in other directions. (Reprinted, with permission, from Fuchs et al. 1981.)

intracellular ions. However, some other cells develop novel electrophysiological properties in culture. When L motor neurons are isolated for several days, their resting potential increases, pronounced rectification with hyperpolarizing current develops, and large amplitude impulses can be evoked in the soma. These changes suggest an increase in excitability of a normally inexcitable cell membrane, similar to those observed following axotomy of insect and crayfish neurons (e.g., Pitman et al. 1972). An increase in input resistance is observed in cultured cells, presumably owing to the detachment of the cell from its

FIGURE 14

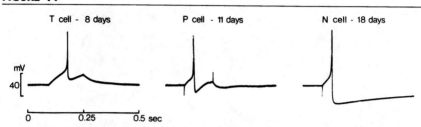

▲ Action potentials of T, P, and N cells after several days in culture. Each cell was from a different culture. The resting potential was −60 mV for the T cell and −50 mV for the P and N cells. The characteristic shape of the action potential in each cell was similar to that recorded from homologous cells in the ganglion. (Reprinted, with permission, from Fuchs et al. 1981.)

processes and from other neurons to which it was coupled in the ganglion.

The electrical synapses seen in culture resemble those in the animal in many respects. The appropriate cells can be electrically coupled, and they have synaptic properties that are normal with respect to rectification. Thus, pairs of R cells placed next to one another in a dish become coupled by nonrectifying junctions by 6 days (Fig. 15). The coupling between R cells can be so strong that a signal generated in one cell is not attenuated at all in the other, although, in general, coupling is less effective that that observed in situ. Since cells are usually placed close to one another to facilitate the formation of connections, coupling ratios reflect the properties of the junction more directly than those measured in situ, where spread of current is attenuated to a variable extent along processes within the neuropil. Coupling also occurs between L cells, and L cells also become coupled to R cells via nonrectifying junctions.

Synapse formation in vitro is not random but instead shows consistent patterns of selectivity. For example, P cells do not become electrically coupled to R cells but they do form rectifying junctions with L cells, as is the case in situ. Chemical synaptic interactions also show specificity: R cells give rise to chemically mediated synaptic potentials in P cells in culture. But the transmission occurs in only one direction; impulses in P cells do not produce synaptic potentials in R cells (Fig. 16).

In culture, the selective formation of connections by isolated neurons depends on inherent properties of the cells and not necessarily on factors such as directed growth, access to synaptic sites, or competi-

FIGURE 15

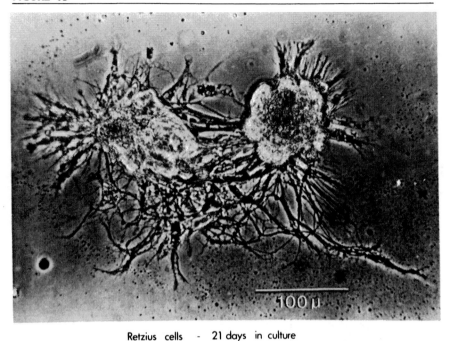

▲ Electrical coupling that developed between a pair of R cells maintained in culture for 21 days is similar to coupling between such cells in the animal. (Reprinted, with permission, from Fuchs et al. 1981.)

FIGURE 16

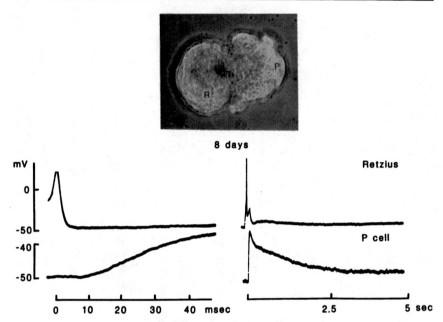

▲ Time course of a reversed IPSP evoked in P cells by impulse in the R cell. The two cells had been cultured together for 8 days. In this record, the electrode in the P cell contained 3 M KCl to reverse and amplify the synaptic potentials, and the external Ca^{++} concentration was 10 mM; no electrical coupling was observed. (Reprinted, with permission, from Fuchs et al. 1981.)

tion. With two or more cells directly in contact and in the absense of a complex neuropil, the chemical and electrical synapses formed in culture appear favorable for testing the ability of a particular cell to form synapses with a wide variety of potential targets and for defining cellular mechanisms that may be involved in specificity of connections.

CONCLUSIONS

In the leech, regeneration of connections has been characterized between individual, functionally identified neurons. The types of synaptic interactions and the morphology of synapses and their modifiability have counterparts in the mammalian CNS, where function and structure cannot be so readily traced to individual cells. We know, however, that no static picture of neuronal connections can be complete. Short-term and long-term changes in synapses occur with neuronal impulse activity or in response to interruption of distant axons. Under certain conditions established synapses between neurons may change permanently. Leech axons can regenerate to form connections similar to those made during development, so that normal functioning is restored. Just how newly formed synapses between particular neurons compare in structure and function with the old, what role in synapse formation is played by the environment or by intrinsic factors, and what cues regenerating neurons use to home in on their proper targets are all problems whose solutions may be at hand in the leech.

REFERENCES

Baylor, D.A. and J.G. Nicholls. 1971. Patterns of regeneration between individual nerve cells in the central nervous system of the leech. *Nature* **232**:268-270.

Calabrese, R.L. 1977. Regeneration of an intersegmental interneuron in the leech. *Neurosci. Abstr.* **3**:296.

Carbonetto, S. and K.J. Muller. 1977. A regenerating neurone in the leech can form an electrical synapse on its severed axon segment. *Nature* **267**:450-452.

Elliott, E.J. and K.J. Muller. 1981. Long-term survival of glial segments during nerve regeneration in the leech. *Brain Res.* **218**:99-114.

Fernández, J.H. and M.S.G. Fernández. 1974. Morphological evidence for an experimentally induced synaptic field. *Nature* **251**:428-430.

Fuchs, P.A., J.G. Nicholls, and D.F. Ready. 1981. Membrane properties and selective connexions of identified leech neurones in culture. *J. Physiol.* **316**:203-223.

Jansen, J.K.S. and J.G. Nicholls. 1972. Regeneration and changes in synaptic connections between individual nerve cells in the central nervous system of the leech. *Proc. Nat. Acad. Sci.* **69**:636-639.

Jansen, J.K.S. and J.G. Nicholls. 1973. Conductance changes, an electrogenic pump and the hyperpolarization of leech neurones following impulses. *J. Physiol.* **229**:635-665.

Jansen, J.K.S., K.J. Muller, and J.G. Nicholls. 1974. Persistent modification of snyaptic interactions between sensory and motor nerve cells following discrete lesions in the central nervous system of the leech. *J. Physiol.* **242**:289-305.

Kandel, E.R. 1976. *Cellular basis of behavior: An introduction to behavioral neurobiology.* W.H. Freeman, San Francisco.

Kristan, W.B., Jr. and P.B. Guthrie. 1977. Acquisition of swimming behavior in chronically isolated single segments of the leech. *Brain Res.* **131**:191-195.

Miyazaki, S. and J.G. Nicholls. 1976. The properties and connexions of nerve cells in leech ganglia maintained in culture. *Proc. R. Soc. Lond. B* **194**:295-311.

Miyazaki, S., J.G. Nicholls, and B.G. Wallace. 1976. Modification and regeneration of synaptic connections in cultured leech ganglia. *Cold Spring Harbor Symp. Quant. Biol.* **40**:483–493.

Muller, K.J. 1979. Synapses between neurones in the central nervous system of the leech. *Biol. Rev.* **54**:99–134.

Muller, K.J. and S. Carbonetto. 1977. Two ways that an electrical connection is re-established in the leech. *Neurosci. Abstr.* **3**:353.

———. 1979. The morphological and physiological properties of a regenerating synapse in the C.N.S. of the leech. *J. Comp. Neurol.* **185**:485–516.

Muller, K.J. and S.A. Scott. 1979. Correct axonal regeneration after target cell removal in the central nervous system of the leech. *Science* **206**:87–89.

———. 1980. Removal of the synaptic target permits terminal sprouting of a mature intact axon. *Nature* **283**:89–90.

Parnas, I., S. Hochstein, and H. Parnas. 1976. Theoretical analysis of parameters leading to frequency modulation along an inhomogeneous axon. *J. Neurophysiol.* **39**:909–923.

Parnas, I., M.E. Spira, R. Werman, and F. Bergman. 1969. Nonhomogeneous conduction in giant axons of the nerve cord of *Periplaneta americana*. *J. Exp. Biol.* **50**:635–649.

Pitman, R.M., C.D. Tweedle, and M.J. Cohen. 1972. Electrical responses of insect central neurons: Augmentation by nerve section or colchicine. *Science* **178**:507–509.

Ramon y Cajal, S. 1928. *Degeneration and regeneration of the nervous system.* (Translated and edited by R.M. May.) Hofner, New York.

Ready, D.F. and J. Nicholls. 1979. Identified neurones isolated from leech CNS make selective connections in culture. *Nature* **281**:67–69.

Sargent, P.B., K.-W. Yau, and J.G. Nicholls. 1977. Extrasynaptic receptors on cell bodies of neurons in central nervous system of the leech. *J. Neurophysiol.* **40**:446–452.

Scott, S.A. and K.J. Muller. 1980. Synapse regeneration and signals for directed axonal growth in the C.N.S. of the leech. *Dev. Biol.* **80**:345–363.

Thompson, W.J. and G.S. Stent. 1976. Neuronal control of heartbeat in the medicinal leech. II. Intersegmental coordination of heart motor neuron activity by heart interneurons. *J. Comp. Physiol.* **III**:281–307.

Van Essen, D.C. 1973. The contribution of membrane hyperpolarization to adaptation and conduction block in sensory neurones of the leech. *J. Physiol.* **230**:509–534.

Van Essen, D.C. and J.K.S. Jansen. 1977. The specificity of re-innervation by identified sensory and motor neurons in the leech. *J. Comp. Neurol.* **171**:433–454.

Wallace, B.G., M.N. Adal, and J.G. Nicholls. 1977. Sprouting and regeneration of synaptic connexions by sensory neurones in leech ganglia maintained in culture. *Proc. R. Soc. Lond. B* **199**:567–585.

Yau, K.-W. 1976. Receptive fields, geometry and conduction block of sensory neurones in the C.N.S. of the leech. *J. Physiol.* **263**:513–538.

Appendix A

Killing Single Cells

Itzchak Parnas
Neurobiology Department
The Hebrew University
Jerusalem, Israel

The nervous system of the leech, with its identified neurons, well-known synaptic interactions, and precisely defined fields if innervation of the body wall, is a convenient preparation for studying changes occurring during regeneration or after injury at the cellular level (Chapters 5 and 10).

In this chapter, it will be shown that it is possible to kill an individual cell in its entirety in a living animal and assess the consequences that follow. Such experiments involving a minimal lesion make it possible to ask questions about (1) the role of an individual cell in a circuit, (2) the way in which one neuron can take over the function of another, and (3) the factors that influence nerve cell sprouting. In the leech, as in several other invertebrates (Hoy et al. 1967), it has been shown that axons can survive without degenerating for periods of days or weeks after they have been severed from their cell bodies. Not only do these processes survive, but through their synaptic interactions in the neuropil, they continue to serve as functional links in their normal pathways. Thus, the technique of severing an axon from its soma to induce its degeneration, which is a key technique in vertebrates, is of no use in the leech in experiments that require complete degeneration. Thus, it was necessary to develop new techniques to destroy a neuron including all of its processes, namely by injecting toxic materials into its cell body through a microelectrode.

But how can one be sure that the injected cell and all of its processes have been killed? For example, intracellular injection of cobalt produces rapid electrical death—but a few days later, the cell can recover and regain its activity. In view of such uncertainties and to be sure that a neuron is dead, it and its arborization should be destroyed physically. The technique to be described here consists of injecting a material that diffuses throughout the cell and lyses it (Bowling et al. 1978).

Ideally, the material used for injections should be active intracellularly and should spread throughout the arborization of the cell. At the same time, the material should be relatively innocuous when applied extracellularly, so that killing one cell would not damage other neurons. The material should be degradable and the animal should be able to inactivate it or secrete it. In addition, it would be convenient for assessing damage if the extent of this material's spread into the small terminals could be observed visually.

These requirements suggest the use of proteolytic enzymes, which are active within the cells and inactivated by protease inhibitors present in the sera of many animals. In fact, during "natural death" of an injured cell, proteolytic enzymes released from lysosomes actually lyse the cell.

INJECTION TECHNIQUE

A commercially available mixture of proteolytic enzymes (Sigma, Protease Type VIII) is suitable for selective and complete destruction of the injected cell. The enzyme mixture reacts optimally at 37°C, and its activity can be greatly slowed by cooling the animal to 4°C, a temperature at which the leech can live for many days. Thus, it is possible to inject the protease into a cell and, after a desired period, stop its action by cooling. Small amounts of the protease are sufficient to destroy a neuron selectively when injected intracellularly. The same amounts, and even larger amounts, do not kill when injected extracellularly.

The microelectrodes used for injection contain a solution of 0.5% protease, 0.4% Fast Green (which enables one to follow the injection visually), and 50 mM KCl (to lower electrical resistance). During good injections, the resting potential and electrical activity of the penetrated cell change little. The cell remains green for about 30 minutes, thereafter it loses the color, presumably because of dye leakage. Cells injected with only 0.4% Fast Green and 50 mM KCl can remain green for 3 days or more.

SPREAD THROUGH CELLS AND ABSENCE OF DAMAGE TO OTHER NEURONS

The lack of damage to neurons other than the injected one has been shown in several control experiments:

1. Injection of protease into one of the two electrically coupled Retzius (R) cells destroys the injected cell without any measurable effects on the second cell.

2. Killing several touch (T), pressure (P), or nociceptive (N) mechanosensory cells that make synapses on a single postsynaptic motor neuron, the longitudinal (L) cell, does not change synaptic potentials produced by the remaining N cell.

3. Extracellular injection of protease produces no obvious changes in the structure or function of cells in the ganglion.

4. Studies by electron microscopy (Bowling et al. 1978; Scott and Muller 1980) show that the space vacated by the killed cell is filled by other neuronal processes, glia, and microglia and that their appearance is normal.

The spread of the protease through the cell can be seen by microscopy as shown in Figure 1. Here, two T cells were injected with horseradish peroxidase (HRP); then 3 hours later one of the cells was injected with the protease mixture and the other with Fast Green and KCl alone. The protease hydrolyzes the HRP, so that the absence of the staining reaction and its product indicates the spread of protease. In this way, it is possible to show that the proteolytic enzyme reaches axon terminals.

KILLING SINGLE CELLS IN THE ANIMAL

The control experiments described above have been carried out in isolated ganglia maintained for several days in organ-culture medium (Wallace et al. 1977) and also by injecting protease into neurons within the animals. To inject cells in the animal, the leech first is anesthetized with chlorobutanol (0.15%). The body segment to be opened is elevated by pinning the leech over a small rod; and a small incision is made in the ventral midline to allow the ganglion to emerge from the body wall. A small incision in the ganglion sheath and proper alignment of a light guide permit clear recognition of the cells. After the injection, there is no need to suture the incised segment, since muscle contractions of the body wall prevent bleeding. Healing of the wound

FIGURE 1

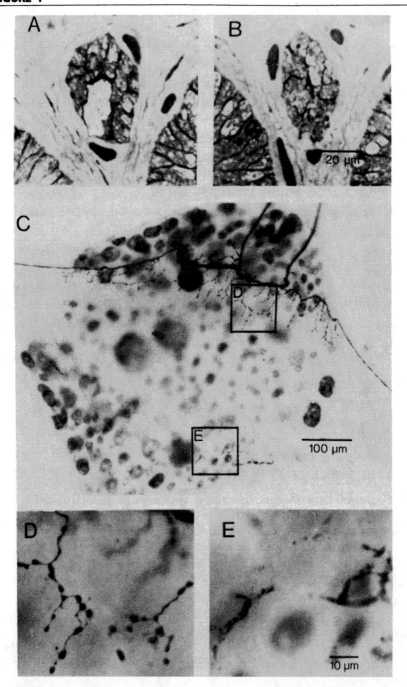

is often so effective that after 2 months it is difficult to find the site of the incision.

The effectiveness of cell killing is shown in Figure 2. Two ipsilateral annulus erector (AE) cells were injected in successive ganglia. As expected from the known fields of innervation (Stuart 1970), the annulus erection reflex disappeared from six annuli. Weeks later, visual inspection and intracellular recordings showed that the two AE cells were indeed missing from the two ganglia. This experiment confirms that the AE cell is the motor neuron responsible for erecting annuli.

The technique of killing neurons by protease injection has been shown to be useful in addressing the following problems.

Circuit Analysis

In the leech, ipsilateral and contralateral N neurons produce synaptic potentials in an L cell (Nicholls and Purves 1970). Electrical recordings cannot reveal whether each N cell makes synapses directly on both L cells or whether the synaptic responses spread from one L cell to the other L cell by way of electrical coupling. Eliminations of one cell with protease has shown that the remaining L cell is innervated by both ipsilateral and contralateral N cells (Bowling et al. 1978).

The T and S cells are coupled electrically by a rectifying synapse that, according to conventional criteria, is characterized as monosynaptic. Recently, Muller and Scott (1981) showed that this coupling takes place through an interneuron. By the use of protease, this neuron was shown to be essential for the coupling between the T and S cells (see Chapter 6 and Appendix D).

Changes in the CNS

The idea that elimination of neurons in the CNS can serve as a signal to other neurons to compensate for or respond to the loss has been tested in several experiments. Significant changes in the central ner-

◀ (A) Cross section of Faivre's nerve. The large axon is that of the S cell. (B) After protease injection, the large axon is missing. (C–E) Double HRP-protease injection experiment. Two T cells were injected with HRP; 3 hours later, one T cell was injected with protease. Note that in C (bottom) the T cell does not show the HRP staining, indicating its destruction by the protease. Higher magnification shows that the terminals from the protease-injected side (E) are disrupted whereas the control (D) looks normal.

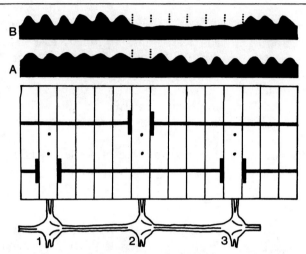

▲ Effect of elimination of one or two AE neurons. Each AE cell innervates nine annuli but not the central annulus of adjacent segments. (A) One AE neuron was killed, and, as expected, one annulus from the fields of innervation as given below failed to erect. (B) When two AE cells in adjacent ganglia were killed, the reflex was lost in six annuli.

vous system (CNS) of the leech have been demonstrated following the destruction of S cells by Scott and Muller (1980) (see Chapter 10). Killing an individual S cell is not sufficient to produce sprouting of the S cells anterior or posterior to the deleted cell. But, if an adjacent S cell is injured at some distant process, it sprouts at the injured end and also at the uninjured tip into the space vacated by the killed cell.

Changes in the Periphery

The precise borders of the fields of innervation of the skin or body muscles of sensory and motor neurons have provided an opportunity for studying changes in the periphery after damage or elimination of cells. Van Essen and Jansen (1977) and Fett (1978) showed that cutting the roots produced little or no spread of T-cell fields from adjacent segments into the deafferented skin.

Slow changes in the periphery are seen in the leech after two ipsilateral AE cells have been killed in two adjacent ganglia. The result is loss of the annulus erection reflex in six annuli. Several months later, the animal regains the reflex. This recovery occurs because adjacent AE cells, on the same side, come to control the reflex in the previously denervated area. It is not clear from this experiment whether the anterior and posterior AE cells indeed sprout into the denervated muscles or whether weak, ineffective innervation exists in normal animals and that such connections become stronger after the major inputs are eliminated.

Sensory neurons offer another convenient system for studying expansion of fields of innervation. For example, killing of three N cells in a ganglion, leaving only one lateral N cell intact, produces expansion of the receptive field of this remaining cell across the dorsal and ventral midline, allowing it to innervate almost the whole circumference of the leech (see Chapter 5). However, killing the two medial N cells, leaving one N cell on each side, does not produce expansion of fields. Killing of N cells does not affect the fields of the T cells in the same ganglion; their fields remain normal. Similarly, killing three T cells on one side produces expansion of only the dorsal skin T neuron across the dorsal midline (never seen in controls); other T cells keep their normal fields and N cells show no expansion of fields.

This finding suggests that the signal to produce these changes is cell specific and that a vacant area is required for homologous cells to expand. Here too, it is not clear whether true sprouting occurs or whether preexisting but inactive branches become active.

Protease injection has been shown to be valuable for deleting single cells in embryos to observe changes that occur subsequently in the embryo (Weisblat et al. 1980). It is also useful in the lobster neuromuscular system where killing of the inhibitory axon produces changes in the excitatory synapses innervating the same muscles (Dudel et al. 1980).

Other techniques of killing single cells or portions of single cells in isolated preparations have been developed by Miller and Selverston (1979). After injection of Lucifer Yellow or other dyes and exposure to strong illumination, the ensuing photochemical reaction kills the cell or axon. This technique promises to be especially useful in isolated preparations for killing one branch of a cell without damaging the rest of it.

REFERENCES

Bowling, D., J. Nicholls, and I. Parnas. 1978. Destruction of a single cell in the central nervous system of the leech as a means of analyzing its connexions and functional role. *J. Physiol.* **282**:169–180.

Dudel, J., Y. Grossman, and I. Parnas. 1980. Synaptic transmission in crustacean muscle; effects of elimination of the inhibitor fiber on excitatory transmission. In *Amino acid transmitters* (ed. P. Mandel and F.V. Defeudis). Raven Press, New York.

Fett, M.J. 1978. Quantitative mapping of cutaneous receptive field, in normal and operated leeches, *Linnobdella. J. Exp. Biol.* **76**:167–179.

Hoy, R.R., G.D. Bittner, and D. Kennedy. 1967. Regeneration of crustacean motoneurons: Evidence of axonal fusion. *Science* **156**:251–252.

Miller, J.P. and A. Selverston. 1979. Rapid killing of single neurons by irradiation of intracellularly injected dye. *Science* **206**:702–704.

Muller, K.J. and S.A. Scott. 1981. Transmission of a 'direct' electrical connexion mediated by an interneurone in leech. *J. Physiol* **311**:565–583.

Nicholls, J.G. and D. Purves. 1970. Monosynaptic chemical and electrical connexions between sensory and motor cells in the central nervous system of the leech. *J. Physiol.* **209**:647–667.

Scott, S.A. and K.J. Muller. 1980. Synapse regeneration and signals for directed axonal growth in the C.N.S. of the leech. *Dev. Biol.* **80**:345–363.

Stuart, A.E. 1970. Physiological and morphological properties of motoneurones in the central nervous system of the leech. *J. Physiol.* **209**:627–646.

Van Essen, D.C. and J.K.S. Jansen. 1977. The specificity of re-innervation by identified sensory and motor neurons in the leech. *J. Comp. Neurol.* **171**:433–454.

Wallace, B.G., M. Adal, and J.G. Nicholls. 1977. Regeneration of synaptic connexions of sensory neurones in leech ganglia in culture. *Proc. R. Soc. Lond.* B **199**:567–585.

Weisblat, D.A., S.L. Zackson, S.S. Blair, and J.D. Young. 1980. Cell lineage analysis by intracellular injection of fluorescent tracers. *Science* **209**:1538–1541.

Appendix B

Immunological Identification of Specific Neurons

Birgit Zipser, Susan Hockfield, and Ronald McKay
Cold Spring Harbor Laboratory
Cold Spring Harbor, New York 11724

Neurons function through a vast number of precise connections that are made with large sets of other neurons; however, very little is known about molecular mechanisms that govern synapse formation. Many hypotheses postulate the presence of specific molecules that mediate cell-to-cell recognition. With the development of the hybridoma, or monoclonal antibody, technology (Köhler and Milstein 1975), it is now possible to search for such unknown molecules within tissue containing a complex mixture of antigens. Neural tissue has such a mixture, including some of the specific molecules synthesized by nerve cells—neurotransmitters and their respective receptors, peptides, and enzymes of synthesis and degradation. Do these classes of molecules constitute the entire range of molecular determinants of neuronal specificity, or are there additional classes of molecules that determine neuronal specificity and that could serve as recognition factors among neurons? The hybridoma technology can now be used to begin to address these questions (Zipser and McKay 1981).

MATERIALS AND METHODS

Monoclonal antibodies were generated by immunizing mice with minced whole leech nerve cords (*Haemopis marmorata*). The spleen cells from these immunized mice were fused with myeloma cells, thereby creating hybridomas that both secreted antibodies and multiplied in tissue culture. These hybridomas were cloned to obtain cell lines that secreted only one specific antibody into the tissue culture supernatant. Leech nerve cords were dissected out, pinned out in a Sylgard dish, and fixed either in Bouin's or paraformaldehyde. To allow antibodies to penetrate across the entire thickness of a ganglion, the ganglion connective tissue capsule was removed with forceps, and the tissue was xylene-extracted; all aqueous solutions contained 0.2% or 2% Triton X (a detergent). The tissue was incubated overnight with a monoclonal antibody, and the antibody was visualized using a goat anti-mouse IgG to which either rhodamine or horseradish peroxidase (HRP) was conjugated.

Of the 475 hybridomas screened, about 300 of them made antibodies that bound to leech central nervous system (CNS) tissue. The majority of these bound to the entire CNS tissue generally. Of special interest to us are those 5–10% of the hybridomas that secrete antibodies specific for small subsets of neurons.

ANATOMICAL MAPPING OF ANTIGENICALLY RELATED NEURONS

Each of the antibodies used recognizes a subset of neuronal cell bodies within the leech CNS. Antibody labeling of nerve cells can serve to elucidate the neuroanatomical organization of the leech nerve cord, with its 34 ganglia and connectives, by demonstrating antigenically homogeneous classes of neurons. Whereas some antibodies recognize previously (or concurrently) identified neurons, such as mechanosensory or motor neurons, others recognize neurons that have not been previously distinguished as a class of related neurons.

Antibody Lan3-9 binds to a group of five cell bodies on each side of the supraesophageal ganglion (Fig. lA). The subesophageal, segmental, and tail ganglia do not contain cell bodies recognized by the antibody. However, antigenically related processes are ubiquitous. Prominent axons run in the connectives along the entire extent of the nerve cord. Occasionally, finer processes are seen in the roots as well. Antigenically related varicosities resembling nerve terminals are abundant in the neuropil (Fig. 1 B and C). An interpretation of this Lan3-9 staining pattern is that cell bodies from the supraesophageal neurons send their processes throughout the entire nerve cord to exert a descending influence on segmental neurons. A first step in proving such a hypothesis will be to determine the actual continuity between antigenically related processes and cell bodies.

Monoclonal antibodies can be used to screen the entire leech CNS in parallel for cell bodies carrying the same antigenic determinant. However, mapping cell bodies immunocytochemically has its limitations. A subset of functionally homologous neurons may lack its class-specific antigen and therefore not be detected. A given antibody might recognize a certain antigenic determinant on two otherwise different molecules, and a set of heterologous neurons that are labeled by a particular antibody would be mistakenly interpreted to share a class-specific antigen.

Sets of morphologically heterologous neurons labeled by two other antibodies are illustrated in Figure 2 B–C. The typical segmental ganglion staining pattern of Lan3-5 includes 10 cell bodies (Fig. 2C) and that of Lan3-6 about 30 (Fig. 2B). Both of these different staining patterns overlap in their staining of mechanosensory pressure (P) cells, indicating the existence of two different class-specific molecular moieties in P cells. This provides an opportunity to map the P-cell distribution that relies not only on one but on two P-cell-specific molecules. Both antibodies verify the normal complement of four P cell bodies, with their right and left dorsal and ventral receptive fields, for segmental ganglia 1–19. The last two unfused segmental ganglia anterior to the fused tail ganglion deviate from this pattern. Instead of binding to four cell bodies, both monoclonal antibodies bind to only the bilateral cell body pair. These ganglia have only 350 instead of the standard 400 neurons (E. Macagno, pers. comm.). Thus, the missing P cell bodies could account for two of the 50 fewer cell bodies. However, the question again arises as to whether these posterior ganglia indeed have fewer P cell bodies or whether all four cell bodies are present, with two of them lacking both molecules specific for P cells.

The nociceptive (N) mechanosensory neuron is identified by monoclonal antibody Lan3-2. As seen in Figure 2D, in most segmental ganglia Lan3-2 binds to four N cell bodies. Again, similar to the variation in P-cell distribution, ganglia 20 and 21 each have only one pair of N cell bodies. Curiously, ganglia 5 and 6, which are specialized to subserve reproductive function, lack Lan3-2–labeled N cell bodies altogether. The head (subesophageal) ganglion has only two pairs of cell bodies binding Lan3-2. A pair of ventral cell bodies and a smaller pair of dorsally situated cell bodies often occur in the largest, most posterior subganglion of the head ganglion. A similar N-cell distribution has been reported in the head ganglion of *Hirudo* on the basis of electrophysiological studies (Yau 1976).

FIGURE 1

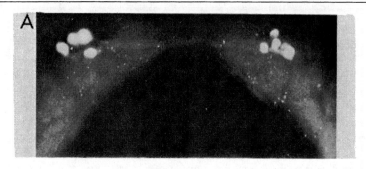

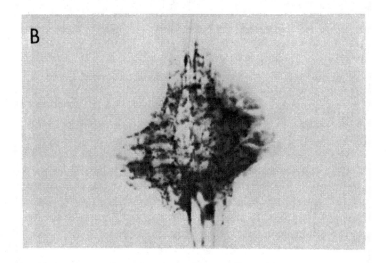

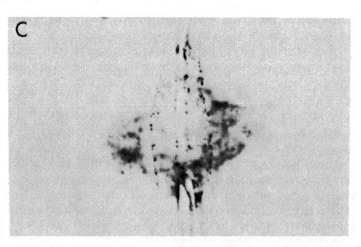

ELECTROPHYSIOLOGICAL CHARACTERIZATION OF ANTIBODY-IDENTIFIED NEURONS

The distribution of mechanosensory P and N neurons provides examples of how monoclonal antibodies can serve as a tool to map variations in the distribution of functionally and chemically homologous cell bodies along the entire extent of the leech nerve cord. However, many antibodies bind to sets of functionally heterologous neurons and can be used to probe into deeper levels of neural organization. Both the antibodies binding to P cells identify other whole subgroups of neurons that share the two antigens. The following questions arise: What is the functional significance of a shared antigen? Does it represent a common neurotransmitter; could it relate to such phenomena as the mechanisms that generate and maintain specific networks? Sexual behavior provides an example of a network that has been analyzed with electrophysiological techniques.

Populations of sex neurons have been identified in ganglia 5 and 6, using nerve backfill with HRP (Zipser 1980) and electrophysiological recording (Zipser 1979a,b). One of the approximately ten cell bodies backfilled in each ganglion is identified by the monoclonal antibody Lan3-1. In ganglion 6, Lan3-1 binds to a pair of 60-μm cell bodies that lie between anterior and posterior roots on the ventral surface of the ganglion (Fig. 3C). With an HRP-conjugated second antibody, one can often see the primary axons of these cell bodies project across the ganglion, giving off a collateral into the contralateral anterior connective before exiting into the contralateral anterior root. This morphology is typical of the lateral PE motor neuron, one of the two major penile evertor (PE) cells.

To demonstrate unequivocally that the large cells stained by Lan3-1 in ganglion 6 were the previously identified PE neurons, a double-labeling experiment using an injected dye was carried out. A cell body believed to be the left lateral PE cell was impaled with a microelectrode and its identity confirmed by its characteristic action potential (20 mV,

◄ Specific staining with Lan3-9. (A) This antibody recognizes a bilateral set of five cell bodies in the supraesophageal ganglion using a rhodamine-conjugated second antibody. These differ from those stained by Lan3-1 (Fig. 3A). (B, C) Two focal planes through the same whole midbody ganglion stained by the peroxidase-antiperoxidase (PAP) procedure. No cell bodies can be seen here, but several axons marked by discontinuous staining and a complex dense array of varicosities are visible. Anterior at top.

FIGURE 2

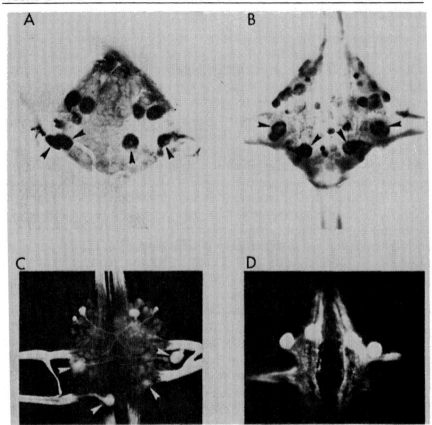

▲ Mechanosensory cells stained by four different antibodies. Three antibodies, Lan3-7 (*A*), Lan3-6 (*B*), and Lan3-5 (*C*), stain the P cells (▲) as part of three distinct staining patterns. The P cells were identified by their morphology and, in the cases of Lan3-5 and Lan 3-6, by double labeling with rhodamine-conjugated second antibody and Lucifer Yellow microinjection. (*D*) The antibody Lan3-2 is shown staining the N cell bodies and processes. Anterior at top.

nonovershooting) and its synaptic relationship with another major PE cell, the rostral PE motor neuron (Zipser 1979). Following electrophysiological identification, the fluorescent dye Lucifer Yellow (Stewart 1978) was injected through the recording electrode (Fig. 3D). To control for the effect of Lucifer Yellow on antibody staining, the left Retzius (R) cell in the center of the ganglion was also injected. After the ganglion was fixed, it was incubated in hybridoma supernatant; antibody binding was visualized with rhodamine-conjugated rabbit anti-mouse IgG. The results illustrated in Fig. 3C and D confirm that the lateral PE cell contains the antigen to which the monoclonal antibody Lan3-1 binds. The same cell fluoresced yellow with the intrasomatically applied marker Lucifer Yellow (Fig. 3D) and red with the rhodamine-conjugated anti-mouse antibody (Fig. 3C). The monoclonal antibody bound to both the left injected lateral PE and the right uninjected cell. When the preparation is viewed under fluorescence filters suitable for observing Lucifer Yellow fluorescence, low intensity of rhodamine fluorescence is seen in the right lateral PE cell. However, Lucifer Yellow fluorescence does not interfere with the antibody visualization, since under fluorescence filters suitable for viewing rhodamine, the R cell shows no fluorescence.

The large Lan3-1–labeled cell bodies in ganglion 5 also lie between the anterior and posterior roots on the ventral aspect. They were previously identified as sex neurons by backfilling a sex nerve with HRP. Their contralaterally projecting primary axons and nonovershooting somatic action potential are properties shared with motor neurons. It is already clear that this pair of presumptive sex motor neurons in ganglion 5 does not elicit penile eversion, although its specific function still needs to be determined.

Thus, Lan3-1 binds to two functionally different neurons within the reproductive network. Electrophysiological evidence indicates that the two types of Lan3-1–labeled neurons belong to a synaptically connected subnetwork within the larger sex network. The ganglion-5 and -6 neurons are linked to each other through a synaptic connection that is mutually excitatory. Stimulating either one of these neurons with a long-lasting depolarizing pulse that evokes a train of spikes generates a distinct series of excitatory postsynaptic potentials (EPSPS) in the other one. To establish whether the antigenic determinant Lan3-1 specifically relates to sexual function, it is necessary to get similar electrophysiological characterizations on the remaining neurons in the staining pattern.

The other cell types labeled by Lan3-1 are illustrated in Figure 3 and in the diagram of the whole leech nerve cord (Fig. 4). Each free segmental ganglion possesses one pair of smaller cell bodies situated dorsally (Fig. 3B). Similar, small-diameter neurons are also found in some of the subganglia in the head and tail brain. In addition, larger-diameter neurons occur in the supraesophageal ganglion (Fig. 3A). Our previous electrophysiological experiments have detected large neurons in the supraesophageal ganglion that could initiate genital motion. Double-labeling experiments are needed to determine wheth-

FIGURE 3

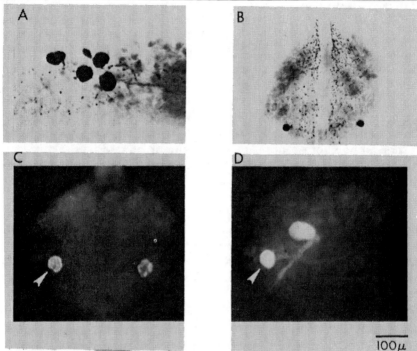

▲ Specific staining with Lan3-1. The monoclonal antibody was visualized immunocytochemically using peroxidase-conjugated (*A,B*) or rhodamine-conjugated (*C*) second antibody. (*A*) Cell bodies in the right supraesophageal ganglion stained with Lan 3-1. (*B*) A bilaterally symmetrical pair of reactive cell bodies in a typical midbody ganglion. In additon to the two deeply stained cell bodies, the neuropil contains a large number of stained beaded processes (varicosities) that extend into one or more axons in the connective. (*C*) Two larger cell bodies that occur reproducibly in the fifth and sixth ganglia stained with a rhodamine-conjugated second antibody. The two smaller cell bodies are also present, but they lie in a different focal plane. (*D*) The same ganglion as in *C* but viewed under fluorescein isothiocyanate (FITC) optics, which reveal the presence of the microinjected fluorescent dye Lucifer Yellow. The left lateral PE cell (▷) was identified by a unique synaptic relationship to the morphologically and physiologically identifiable rostral PE cell. The other cell labeled by Lucifer Yellow as a control is the R cell. Anterior at top.

er these sex-related neurons are identical to those that carry the Lan3-1 antigen.

FIGURE 4

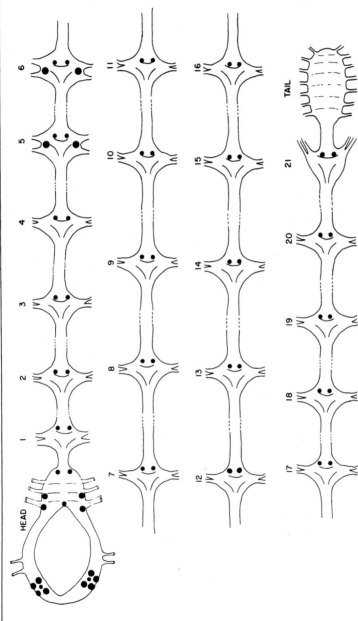

▲ This diagram illustrates the symmetry and repetitive organization of the nerve cord. The cell bodies stained by Lan3-1 in the 21 free midbody ganglia were verified in a large number of tests. The head and tail brains are more difficult to analyze and might contain additional neurons besides those indicated.

In addition to Lan3-1, there is another antibody (Lan3-3) that only binds to supernumerary cells in the sex ganglia. However, because the Lan3-3 neurons are not yet identified electrophysiologically, it is not known whether they are related to sexual function. What is particularly interesting is that, in addition to the large supernumerary cells in ganglia 5 and 6, Lan3-3, just like Lan3-1, labels a pair of smaller cell bodies that repeat in all segmental ganglia.

ANATOMICAL MAPPING OF ANTIGENICALLY DEFINED AXONS

Monoclonal antibodies have been shown to differentiate subpopulations of neurons. In most cases, the antibodies stain neuronal cell bodies and their processes. Therefore, one can use monoclonal antibodies to study the organization of the connective. The connectives between each pair of segmental ganglia contain approximately 5000 axons, a large number relative to the 350–400 neurons in each ganglion. The origin and organization of these axons is largely unknown. Do the axons occupy stereotypical positions in the connectives as neuronal cell bodies occupy stereotypical positions in the ganglia; do axons, like cell bodies, carry specific molecules?

Connectives from fixed leech nerve cords were embedded in gelatin and cross-sectioned on a vibratome at 150 μm. Serial sections were processed for different antibodies. Some sections were postfixed in OsO_4, embedded in Epon, and sectioned at 1 μm or thin sectioned for electron microscopy.

Figure 5 illustrates staining of the connective using three different monoclonal antibodies. All the sections are from the connective between ganglia 15 and 16. The positions of stained axons are essentially constant in the connective between each pair of ganglia throughout the length of the nerve cord and are consistent among leeches of the same species. Each antibody has a recognizable staining pattern in the connective.

Lan3-9, an antibody that stains ten cell bodies in the supraesophageal ganglion (Fig. 1), stains axons in the medial bundle of axons in the connective (Faivre's nerve) and a few axons in the medial part of each lateral connective (Fig. 5A). Preliminary studies by electron microscopy show that Lan3-9–stained axons have relatively large diameters, that is, larger than the majority of axons in the connective but significantly smaller than the large axon of the S cell in Faivre's nerve.

Lan3-6, an antibody that stains the P cells and other, unidentified neurons (Fig. 2C), stains a bundle of axons in the mediolateral part of

FIGURE 5

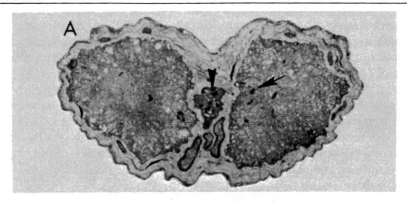

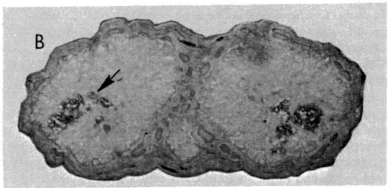

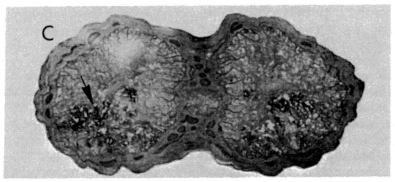

▲ Cross sections of the connectives between ganglia 15 and 16 stained with monoclonal antibodies. (A) Lan3-9 stains axons in Faivre's nerve (▶) and in the medial part of each lateral connective (→). (B) Lan3-6 stains bundles of axons in the mediolateral part of each lateral connective (→). (C) Lan3-2 stains several bundles of axons in the ventrolateral part of each lateral connective (→). Dorsal at top. Magnification, 2000×.

each lateral connective (Fig. 5B). Axons are stained neither in Faivre's nerve nor in the medial part of either lateral connective.

Lan3-2, an antibody that stains N cells (Fig. 2D), stains axons in the ventrolateral part of each lateral connective (Fig. 5C). Whereas the distribution of Lan3-2–stained axons may overlap in some measure that of Lan3-6–stained axons, the overall staining patterns differ sufficiently to distinguish between the two. In preliminary studies by electron microscopy, the range of axonal diameters of Lan3-2–stained axons includes the smallest axons in the connective, whereas the range of Lan3-6–stained axons is somewhat larger. Studies by electron microscopy have verified that Lan3-6– and Lan3-2–stained axons travel in bundles and Lan3-9–stained axons travel singly through the connective.

At present, the quality of tissue prepared for electron microscopy using monoclonal antibodies is quite poor, owing to the use of detergents and solvents and the destructive effect of glutaraldehyde fixation on antigenicity. Using the technology currently available, it has been possible to determine (1) that monoclonal antibody staining can be observed at the level of electron microscopy; (2) approximate diameters for antibody-stained axons; and (3) that entire groups of axons (fascicles) are labeled by some antibodies (Lan3-6 and Lan3-2). Further technological developments will make it possible to determine axon diameters precisely and accurately count the number of axons stained by a particular monoclonal antibody. Correlations between numbers of stained axons and numbers of stained cell bodies may elucidate patterns of neuronal circuitry.

SUMMARY

This chapter has described three approaches to the analysis of the leech nervous system by using specific antibody binding. First, for each antibody one can define the distribution of its binding sites along the whole length of the nerve cord. These maps of antigenically related neurons give us a new view of the organization of the leech CNS. Second, to explore the significance of a particular staining pattern, one can obtain an electrophysiological characterization of neurons sharing a common antigen. Third, one can examine by light and electron microscopy the organization of the connective and reveal subsets of axons that form identifiable fiber tracts throughout the leech nerve cord.

In conclusion, antibodies specific for single types and small sets of neurons have been generated. These antibodies reveal novel principles of organization of the leech CNS. The functions of the antigens have not yet been determined, but the distribution of the antigens suggests that some may recognize sets of neurons sharing common functions and possibly common transmitter molecules. The highly specific and reproducible binding of the antibodies suggests that their antigens could serve as factors for neural recognition.

REFERENCES

Köhler, G. and C. Milstein. 1975. Continuous cultures of fused cells secreting antibodies of predefined specificity. *Nature* **256**:495–497.

Stewart, W.W. 1978. Intracellular marking of neurons with a highly fluorescent naphthalimide dye. *Cell* **14**:741–759.

Yau, K.-W. 1976. Physiological properties and receptive fields of mechanosensory neurones in the head ganglion of the leech: Comparison with homologous cells in segmental ganglia. *J. Physiol.* **263**:489–512.

Zipser, B. 1979a. Identifiable neurons controlling penile eversion in the leech. *J. Neurophysiol.* **42**:455–464.

———. 1979b. Voltage-modulated membrane resistance in coupled leech neurons. *J. Neurophysiol.* **42**:465–475.

———. 1980. Horseradish peroxidase nerve backfilling in leech. *Brain Res.* **182**:441–445.

Zipser, B. and R. McKay. 1981. Monoclonal antibodies distinguish identifiable neurones in the leech. *Nature* **289**:549–554.

Appendix C

The Nervous System of the Leech: A Laboratory Manual

This manual is intended for a 3-week course. The purpose of the course is to teach by use of the relatively simple nervous system of a single animal, a variety of neurobiological techniques and to provide a basis for discussions of such topics as neural signaling, synaptic transmission, integrative mechanisms, and long-term changes in the function of the nervous system.

LEECHES

Sources

Hirudo medicinalis and *Macrobdella decora* can be obtained from European and American suppliers, respectively. Possible sources are listed in Table 1.

In most places, an abundant supply can be collected locally, especially in spring and summer. The ideal place for collecting leeches is a shallow, permanent pond that is rich in aquatic vegetation and snails.

This Appendix is a revised version of the manual *The Nervous System of the Leech* written by Jan Jansen, Kenneth Muller, John Nicholls, and Roy Sawyer and used at Cold Spring Harbor Laboratory.

Appendix

The best way to collect leeches is to wade along the edge of the water, examining submerged objects by hand. The most likely objects to harbor leeches are smooth, man-made surfaces, such as soft drink bottles, metal cans, and pieces of plastic; such surfaces are well suited for attachment and locomotion via suckers. An illustrated key to identification of leech species can be found in Sawyer (1972).

Maintenance

Immediately after collection or receipt of shipment, place the leeches into rain or pond water and maintain them at room temperature (20–25°C) for several days. Never keep leeches in distilled water, which will cause ionic depletion and osmotic stress, nor in tap water (the chlorine added to most urban water supplies will kill or sicken most species). Tap water that has been dechlorinated by aeration for 24 hours can be safely used. The following recipe for ionically balanced artificial pond water using either distilled or deionized (not tap) water is practical for laboratory use:

TABLE 1

Sources for obtaining leeches

Hirudo medicinalis may be obtained from:
Frau E. Nell
Blutegelimport und Versand
4350 Recklinghausen-Sud
Elisabethstrasse 16a, West Germany Telephone 62632

M. Desbarax
RICARIMPEX
33980 Audenge, Gde, France

American leeches (e.g., *Macrobdella decora*) may be obtained from:
Carolina Biological Supply Co.
Burlington, North Carolina 27215
919-584-0381

Carolina Biological Supply Co.
Powell Laboratories Division
Gladstone, Oregon 97027
503-656-1641

Connecticut Valley Biological Supply Co., Inc.
Valley Road
Southampton, Massachusetts 01073

Basic stock solution per liter: 2.8 g NaCl, 0.05 g KCl, 0.08 g Ca(NO$_3$)$_2$ · 4 H$_2$O, 0.025 g MgSO$_4$ · 7 H$_2$O, 0.56 g Tris buffer (pH to 7.4).

Artificial pond water: Dilute the basic stock solution to 1% with distilled or deionized water.

An alternative artificial pond water: 0.5 g solid artificial sea water per liter distilled water.

As most species are very sensitive to sudden temperature changes, take care to avoid a temperature change of more than 5°C in any 30-minute period.

Strictly aquatic species can be kept in water-filled containers; amphibious species, such as *Hirudo*, *Macrobdella*, and *Haemopis*, are best maintained under the following special conditions. Place 10–15 leeches into a 5-gallon aquarium containing 1 inch of water and a little sand rising to a height of about 2 inches at one end, well out of the water. Use masking tape to secure a lid punched with minute air holes so that the leeches cannot escape. Leeches can live under these conditions at 15°C for weeks without feeding. If it becomes necessary to feed them, it is best done at room temperature, at which they should be kept for at least a week after feeding. Food requirements are species-specific (see Table 2).

GENERAL EXPERIMENTAL PROCEDURE

Required Equipment

The electrophysiological station used in this course consists of a vibration-resistant table covered by a heavy steel plate. Right- and left-handed micromanipulators, microscope lamps, a mirror, and a moveable dark-field condensor are held to the surface by strong magnets. A dissecting scope is attached to the table. Next to the table stands an electronic equipment rack, which holds a two-channel oscilloscope and differential preamplifiers, two negative-capacitance electrometers, a two-channel current stimulator and stimulus isolation units, a two-channel voltage calibrator, and an audio monitor (Fig. 1).

The leech nerve cord preparation from which recordings are to be made is pinned to the bottom of a recording chamber containing physiological saline (Ringer's fluid). The recording chamber rests on a stand held to the steel plate by magnets (Fig. 2). The electrolyte in the glass microelectrode used for intracellular recording from leech neurons is inserted into the tip of one of the two electrometer probes, each of which is, in turn, held by one of the two micromanipulators. Suction electrodes used for taking extracellular recordings from leech

TABLE 2

Classification of the most common leech species along with useful characteristics

	Aquatic or amphibious	Type of development (size of egg)	Laboratory breeding	Food
Order Rhynchobdellida				
Family Piscicolidae (The fish leeches)				
Family Glossiphoniidae				
Helobdella triserialis (= *lineata*)	aquatic	direct (0.5 mm)	easy	snails
Helobdella stagnalis	aquatic	direct (0.5 mm)	difficult	Tubifex
Placobdella parasitica (or *ornata*)	aquatic	direct (1.5 mm)	difficult	turtles
Haementeria ghilianii	aquatic	direct (2.5 mm)	easy	rabbits
Order Gnathobdellida				
Family Hirudinidae				
Hirudo medicinalis	amphibious	albumenotrophic (about 60 μm)	difficult	frogs, rabbits
Haemopis sanguisuga	amphibious	albumenotrophic (about 60 μm)	difficult	earthworms
Macrobdella decora	amphibious	albumenotrophic (about 60 μm)	difficult	frogs
Order Pharyngobdellida				
Family Erpobdellidae				
Mooreobdella microstoma	aquatic	albumenotrophic (about 60 μm)	easy	Tubifex
Erpobdella punctata	aquatic	albumenotrophic (about 60 μm)	difficult	Tubifex

FIGURE 1

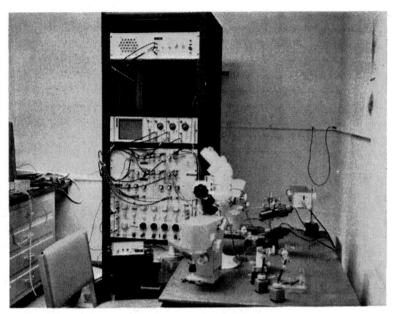

▲ Experimental set-up — electronic equipment.

nerves are held in position by special stands attached to the steel plate.

The following points must be made concerning the electronic equipment in preparation for recording:

1. Equipment manuals should be referred to for the operation of individual pieces of electronic equipment, especially for the oscilloscope, electrometers, and stimulator.

2. The use of a common ground, to which all pieces of apparatus not otherwise grounded are connected by independent direct leads, will reduce 60-cycle electrical noise in the recordings.

3. The oscilloscope and electrometer amplifiers should be adjusted so that a signal of zero volts at the input registers as zero volts at the output.

4. The negative-capacitance electrometers provide a bridge circuit that allows voltage to be monitored while passing current. To gain experience with the use of the bridge circuit, it is a useful exercise to connect a 20 MΩ or higher resistance between the probe input

254 Appendix

FIGURE 2

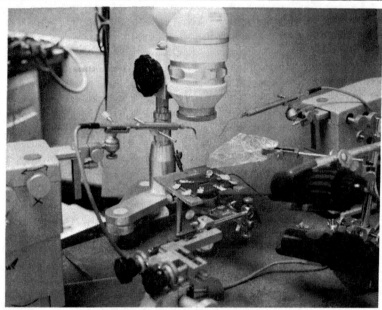

▲ Arrangement of equipment on table for recording.

and ground. Adjust the bridge balance and measure current and voltage while passing current pulses using the stimulator. Observe that the negative capacitance control affects the frequency response, and thus the shape and amplitude of transient voltage changes such as action potentials.

Microelectrodes

Intracellular recordings are taken by means of glass pipette microelectrodes drawn to a very fine and sharp point. The Hockman Puller with two large wheels is the best device for drawing microelectrodes for impaling leech neurons. Use glass capillaries made by Frederick Haer (cat. #30-30-0), 1-mm. O.D., and thin-walled with fiber ultradot. Ideally, the heating foil of the puller should be U-shaped and positioned as close as possible to the capillary. But if the foil is too close, the glass will adhere to and twist the foil. If it is necessary to make a new heating foil, use a strip of platinum 4 mm wide and 2 mils thick. As for the shape of the microelectrodes, a gentle continuous taper with neither convexity nor concavity is best. Examine under a high-power microscope each microelectrode that is made; the tip itself should look fuzzy and be beyond the resolving power of the light microscope (see Fig. 3). To fill the microelectrode, inject the shank with 4 M potassium acetate, using a fine syringe needle. The microelectrode will fill in a few minutes. For best results, wait for 2 hours. Bubbles do not matter. Ignore them.

Dissection of the Ganglia

Two sharp pairs of Dumont No. 5 forceps and two pairs of scissors, one fine and one coarse, are needed. Never use the fine scissors for cutting anything except delicate tissues, such as the nerve cord or the blood sinus surrounding it. Before starting, prepare a small beaker of Ringer's fluid in which to place the dissected ganglion and a squirt bottle of Ringer's fluid to irrigate the preparation (see Table 3 for

FIGURE 3

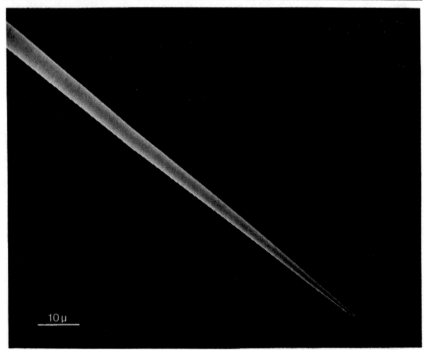

▲ Scanning electron micrograph of electrode. (Photographed by G. Albrecht-Bühler.)

TABLE 3

Leech Ringer's fluid

Solution		One liter (stock in ml/liter)
stock solutions		
NaCl 3 M (M.W. = 58.4; 3 M = 175.2 g/liter)		
KCl 1 M (M.W. = 74.56; 1 M = 74.6 g/liter)		
$CaCl_2$ 1 M (M.W. = 147.06; 1 M = 147.06 g/liter)		
Tris maleate 0.2 M		
To make: 24.2 g Tris + 23.2g maleic acid, add 900 ml H_2O;		
Neutralize to pH 7.4 with 3 N NaOH (100 ml)		
Dilute to 1 liter		
Normal Leech Ringer's fluid		
NaCl	115.0 mM	38.3
$CaCl_2$	1.8 mM	1.8
KCl	4.0 mM	4.0
Tris maleate	10.0 mM (pH 7.4)	50.0
High $CaCl_2$		
$CaCl_2$	116.8 mM	116.8
KCl	4.0 mM	4.0
Tris M	10.0 mM	50.0
NaCl	—	—
8 mM high $CaCl_2$		
$CaCl_2$	8 mM	8
NaCl	115.0 mM	38.3
KCl	4.0 mM	4.0
Tris	10 mM	50
High $MgCl_2$		
$MgCl_2 \cdot 6 H_2O$ (1 m)	116.8 mM	116.8
KCl	4.0 mM	4.0
Tris	10.0 mM	50.0
NaCl	—	—
High NaCl		
NaCl	116.8 mM	38.9
KCl	40.0 mM	4.0
Tris	10.0 mM	50.0
$CaCl_2$	—	—
High KCl		
KCl	119.0 mM	119.0
$CaCl_2$	1.8 mM	1.8
Tris	10.0 mM	50.0
NaCl	—	—

Sulfate Ringer's fluid
Na$_2$SO$_4$ (1 m)	57 mM	57
K$_2$SO$_4$ (1 m)	2 mM	2
CaCl$_2$ (1 m)	1.8 mM	1.8
Sucrose (1 m)	7.6 mM	7.6
Tris maleate	10.0 mM	50

Note: Add CaCl$_2$ after sulfates have been diluted to 1 liter.

15 mM Ca^{++}: 885 ml normal Ringer's + 115 ml high CaCl$_2$

15 mM Mg^{++}: 128 ml high MgCl$_2$ Ringer's + 872 ml normal Ringer's

---

solutions). A heat filter must be placed over the lamp because overheating and drying spoil the preparation during dissection.

Pin out the leech so that it is lying on its dorsal, dark surface. In general, use ganglia from the middle and tail end of the body, and avoid the first six ganglia, unless the heart interneurons are being examined (see Fig. 2 in Chapter 4). Later when removing body wall and skin with ganglia, it may be useful to approach dorsally. Next, find the sensilla that mark the positions of ganglia and cut along the midline longitudinally through the skin and muscles. Notice particularly the oblique crisscross layer of fibers superficial to the longitudinal muscles (see. Fig. 3 in Chapter 4). The nerve cord lies in a green blood sinus (the stocking) in the exact midline. It is important to irrigate and to repin the preparation frequently as the dissection proceeds. The stocking is slit longitudinally, first over and then under the ganglion in the midline. Leave some of the stocking attached to each pair of roots to provide a means of pinning the ganglion. The main dangers in dissecting are: (1) drying out, (2) overheating, and (3) bursting the gut. It is a good idea to take out three or four ganglia for a few hours work. They keep well in Ringer's fluid fortified with a small amount of glucose (approximately 10 mM).

Adjusting the Light

This is very important and should be done before pinning out the ganglion in the recording chamber. First, focus the light on the flat surface of the mirror using a piece of tissue paper. Then tilt the mirror so that light is directed vertically upwards to the recording chamber. Again use a piece of paper to adjust the light and the mirror. Finally, slide the dark-field condenser into place directly above the mirror. It should be oriented so that its surfaces are horizontal with respect to the recording chamber and the steel plate. As it is moved up and down, there comes a point, high up, at which a dark circle surrounded by a light ring appears. This should be concentric. If it is not, tilt the condenser slightly or adjust the mirror. For optimal illumination, the condenser is usually moved up above this level with only a small black spot. Once everything has been adjusted, do not move the mirror or the lamp or tilt the dark field. Simply raise or lower the dark field or move it horizontally for best illumination.

Mounting the Preparation

Unless one wants to record from motor neurons on the dorsal surface, the ganglion should be pinned ventral surface uppermost. Notice that the connectives curl downward when the ventral surface is facing upward. One pin holds each pair of roots by the stocking and each pair of connectives. The anterior part of the ganglion is recognized by the large Retzius (R) cells. Also, if an imaginary line is drawn across the ganglion connecting the roots from side to side, the anterior end is the larger part. The dorsal side of the ganglion can be recognized by the large central area that is devoid of cells. When pinning the ganglion, stretch it gently so that it lies flat. Too great a stretch makes it impossible for microelectrodes to penetrate cells for reasons that are unknown.

An Important Note

Once the ganglion is in the chamber *the fluid should be changed at least once every 10 minutes* or it becomes hypertonic and the cells shrink or the ganglion swells or both.

EXERCISE 1: IMPALING CELLS

The initial exercise is to impale nerve cells in the ganglion. Good visibility and a sharp microelectrode are absolute requirements for success (see Figs. 1 and 4 for set-up). Check the resistance first. If the value is less than 20 MΩ or more than 100 MΩ, then change it; if the value is too high, "ring" it. Use 100× magnification with the dissection microscope. Adjust the dark-field condenser, so that both the microelectrode and the ganglion are seen distinctly.

Select one of the larger cells in the ganglion. Use a time base of 10 msec/div and let it run continuously (i.e., on "line"); use a gain of 10 mV/div. Place the tip of the microelectrode as close as possible to the center of the cell body. Lower the microelectrode with the microdrive, so that the cell just dimples and the circumference of the cell expands. Gentle tapping of the micromanipulator near its base should now be sufficient to penetrate the cell. A good rule is that if the tap can be heard, it is too hard. As the microelectrode penetrates, watch the oscilloscope, not the cell. Successful impalement is demonstrated by recording a resting membrane potential and usually signs of electrical activity (action potentials or synaptic potentials) from the cell. If this first trial does not lead to impalement, advance the microelectrode slightly and try again. If still unsuccessful, replace the microelectrode and try another one.

A common pitfall: If the probe cannot be balanced and all electrodes seem to have infinite resistance, be sure that the bath electrode is in fluid and grounded.

It takes some experience to impale leech neurons well, and the quality of the impalement varies enormously. Only experience teaches recognition of a good impalement, and that is the main object of the exercises over the first few days. Impale one cell after another. Note for each cell:

1. The value of the resting membrane potential. Is it stable?

2. Indications of impulse activity in the cells. Are there overshooting action potentials? What are their amplitudes and time courses?

3. Are there signs of synaptic activity in the record? What are their amplitudes and patterns of appearance? Use the loudspeaker connected to the amplifier. Invariably, a good impalement gives a characteristic sharp click in the loudspeaker. With a little experience, the different types of cells can be recognized from the sound of their action potentials in the loudspeaker.

The main rules for impaling leech neurons are:

1. Optimal visibility.

2. Change the microelectrode at once if there are difficulties.

3. Change the ganglion if it appears opaque or if the cells are shrunken.

4. Remember to change fluid every 10–15 minutes or so.

EXERCISE 2: THE SENSORY CELLS

The next step is to demonstrate that a particular neuron has characteristic and recognizable electrical properties and that there is a constant

FIGURE 4

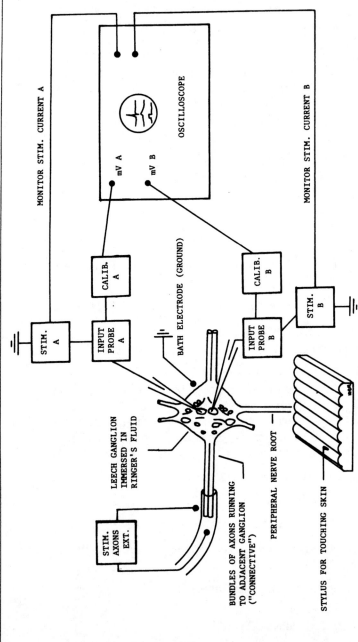

▲ Diagram of arrangement for recording from single cells in a leech ganglion. Stimuli can be applied via the microelectrodes, by external electrodes on the axons of the connective, or by touching the skin.

complement of neurons of each type in all ganglia. The mechanosensory cells are most instructive in this connection and are the object of this exercise. Again, use the isolated ganglion preparation.

There are three different types of mechanosensory cells — pressure (P), touch (T), and nociceptive (N) cells. They are all among the larger neurons in the ganglion. In contrast to most other leech neurons, the sensory cells generate action potentials that overshoot when recorded from the cell body. Each type of cell has an action potential that distinguishes it from the other two types. The functional identification of the cells as mechanosensory neurons belongs to a later exercise.

Mount the isolated ganglion with its ventral side up. Impale large- and medium-sized neurons. Activate the cell by passing an outward current pulse across its membrane. For each cell, start with a weak (10^{-10} A) pulse of 30 msec duration. Balance the bridge of the input amplifier to eliminate the voltage drop across the microelectrode tip. Then increase the current slowly.

Conduct the investigation of the ganglion so that the following questions can be answered:

1. Where are the sensory cells located? Draw a map of the ventral aspect of the ganglion.

2. How many are there of each type?

3. What are the distinguishing features in the action potentials of P, T, and N cells?

4. Activate the cell with long (0.5–1 sec) current pulses and note differences in behavior.

5. Inject subthreshold de- and hyperpolarizing current pulses and note differences in time dependence of response.

6. Monitor the intensity of the current pulses and make a rough estimate of the current-voltage relation (input impedance) of the cell body.

7. Try to activate the neuron at the break of a long (50–200 msec) and relatively large hyperpolarizing pulse (anode break excitation).

During the search for sensory cells, a large number of other neurons in the ganglion will have been impaled. Most of these do not have overshooting action potentials. Their action potentials are small and more or less triangular, usually less than 10 mV in amplitude. That is because the action potential does not invade the membrane of the cell body in these neurons. Some of these neurons will come up again later as the motor neurons. They were designated "pagodas" by Stephen Kuffler because of the resemblance of a train of their action potentials to the roof of a Chinese temple. Try to find the heart excitor (HE) cell described by Thompson and Stent (1976) that controls the leech's heart. Its inhibitory postsynaptic potentials (IPSPs) are quite distinctive.

EXERCISE 3: ELECTRICAL COUPLING

Many of the neurons of the leech ganglion are electrically coupled to each other. Such connections are established through specialized regions of contact between processes of various neurons, and these connections appear to be as invariable and orderly as the chemically mediated synaptic interactions. For the initial exercise, the electrical coupling between the pair of large R cells located in the center of the ventral surface of the ganglion will be studied.

Make sure there are sharp microelectrodes in both micromanipulators. Insert a microelectrode in each of the two R-cell bodies. Note the relationship in the spontaneous firing of the two cells. Pass a weak hyperpolarizing current pulse (200 msec) through one microelectrode and look for changes in the membrane potential of the other cell. Try stronger pulses and pulses of reversed polarity. Pass current through the second microelectrode, look for changes in membrane potential of the first cell.

After the initial casual observations, make a systematic study of the coupling between the two cells.

1. Balance the bridge of the current-passing microelectrode carefully and record the changes in membrane potentials of the two cells for a number of pulses of different intensities. Plot the membrane potential of cell 1 against that of cell 2.

2. Use one amplifier to record membrane current. Plot current against the membrane potentials of cells 1 and 2.

3. Insert both microelectrodes into the same R cell and repeat the same exercises.

It takes some skill to obtain sufficiently stable penetrations to complete all the measurements on one pair of cells. Take a new ganglion when the cells have deteriorated.

How can the degree of coupling between the cells be expressed conveniently? How reliable is the measurement of the input impedance of a cell with a single microelectrode in a bridge circuit?

EXERCISE 4: PATHWAYS OF AXONS DETERMINED BY SUCTION ELECTRODES

It is often useful to map the pathways and distribution of the axonal branches of a particular neuron with electrophysiological techniques. The suction electrode can then be an efficient way of establishing electrical contact with a peripheral nerve or a bundle of axons. The tips of the suction electrodes are prepared by pulling glass tubes to the desired dimension. A good fit of the nerve in the tip of the electrode is required to reduce shunting of the current and obtain optimal performance. Fire polishing of the electrode tip is advisable.

The suction electrode can be used for stimulation of the nerve or for recording its action currents. It is connected through a two-way switch either to the output of the stimulus isolation units or to the A.C. preamplifier.

Dissect isolated ganglia with long stretches of the three large peripheral nerves and long connectives. The nerves must be cleanly dissected without adherent connective tissue or muscle. Separate the two connectives from each other. Pin the ganglion in the usual way, ventral side up with the nerves accessible to the suction electrode. Avoid pinching the nerves to be examined.

Start with the anterior root or connectives in the suction electrode. Use a microelectrode to record the activity of the cells in the ganglion. For each cell, determine first whether it can be activated by stimulation of the nerve and then whether impulses can be recorded from the nerve after stimulation of the cell body by a current pulse. Start with T cells; then try P and N cells. Conduct the investigation so that the following questions can be answered:

1. Which of the sensory cells can be activated by stimulation of the nerve?
2. What is the stimulus intensity required to activate the cells?
3. What are the amplitudes of the impulses recorded from the nerve?
4. Examine synaptic potentials in R cells, T cells, and HE cells. What effect does altering the membrane potential have on excitatory postsynaptic potentials (EPSPs) and IPSPs?
5. Try to find the small unpaired S cell that lies close to an R cell. Its impulse is like that of a T cell and it gives rise to a spike of very large amplitude in the middle connective.

   Examine similarly the axons in the other nerves and the connectives.

Demonstrate impulse collision in an axon by stimulation of both the cell body and the end of the nerve simultaneously. Show that the period of time during which collision occurs is approximately equal to the conduction time of the axon.

How can we know whether an impulse recorded in the cell body is due to direct stimulation of the cell's axon in the nerve? How decisive are the various criteria?

Stimulate the nerve repetitively at 20 Hz, while recording from the various sensory cell bodies. How are the shapes and amplitudes of impulses changed? Try to enhance or counteract the impulse block by changing the membrane potential of the cell body by long-lasting current impulses.

Note: Remember that suction electrodes are far easier to use if a three-way tap is connected to the syringe to prevent the ganglion from being sucked into the syringe or blown out across the room.

EXERCISE 5: THE GANGLION-BODY WALL PREPARATION

In this preparation, the pathways between body wall and the sensory and motor neurons remain intact as required for the functional identification of the different types of neurons.

Perform the standard midline dissection of the nerve cord and ganglia over three to four segments. (Many people prefer a dorsal midline approach to the ventral approach described here.) Select a ganglion and expose the anterior and posterior roots on one side. Identify the central annulus of the segment by its two ventral sensilla. Remove a millimeter or so of the medial skin and the body wall muscles in the three central annuli to increase the lengths of exposed roots. This facilitates the final pinning of the preparation. Cut the contralateral roots outside the blood sinus and cut the two connectives midway towards the neighboring ganglia. The ganglion is then lying closely against the edge of the ventral body wall. Cut the body wall from the ventral midline to half way into the dorsal surface along the central annulus just in front of and just behind the segment to be isolated. Then dissect carefully between the layer of longitudinal muscle and the gut. Some care is required at this stage to avoid injury of the nerves. Cut through the dorsal branch of the posterior root and the bundles of dorsal-ventral flattener muscle. Then, with the gut separated, deflect laterally the ventral part of the body wall and pin it out moderately stretched. Clean the inner side of the longitudinal muscle for what may remain of gut and loose tissue. The ganglion and body wall preparation can now be isolated. The flap contains about eight annuli in the anterior-posterior direction and extends from ventral midline to halfway into the dorsal surface. Now pin it in a Sylgard-coated dish with the skin up and moderately stretched. Use the standard leech Ringer's fluid in the bath. Pull the ganglion clear of the medial edge. Stretch the roots moderately, so that the ganglion is separated from the body wall; pin it ventral side up and avoid twisting the roots. To minimize body wall movements pulling on the ganglion, it should be pinned also between the roots on the body wall side. The sensory cells will be examined first.

A suitable stimulus of mechanoreception is provided by a "toucher" mounted in one of the micromanipulators. The toucher is made from a broken micropipette with its tip flamed to produce a small glass ball. (For finer localization of sensitive spots in the skin and for applying timed touches, use a mechanical stimulator that is driven by current pulses from the stimulator. It consists of a piezoelectric crystal shaped like a thin wafer, which is attached a fine glass stylus made from a broken microelectrode.)

Impale one of the two ipsilateral medial T cells. Touch the various regions of the skin preparation with the toucher. Determine the boundaries of the receptive field of the cell and draw a map of it. It is absolutely necessary to have the signal connected also the loudspeaker. Notice the shape of the impulse on a fast-sweep speed. The rapid, uninterrupted rise of the impulse indicates that the cell is activated directly. Often, a T cell can be activated synaptically from other regions of the skin. The synaptic response is less brisk and the impulse is different from the directly elicited one. Develop a feeling for the intensity of the stimulus required to activate the cell. Is the skin equally sensitive over the entire surface of the receptive field?

Examine the other two ipsilateral T cells, and then the P and N cells in a similar manner. Determine the stimulus intensity required to activate the cells and map their receptive fields. Note that the N cells respond only to very strong stimuli and pinching the skin with a pair of tweezers may be required.

1. Use Ringer's fluid containing 1.8 mM Ca^{++} for this experiment.

2. To make a Sylgard bath, pour a small amount of premixed Sylgard 184 resin into a plastic dish and allow it to cure at about 60°C overnight. The bubbles will disappear.

3. Take care not to let the skin dry out. Change fluid often.

4. If the receptive field of a T cell has been plotted carefully, cut the anterior root and see what happens to the shape of the field. Where is the boundary of the territories innervated by axons running in the anterior and posterior roots?

5. Try to find the additional fields of T cells in adjacent segments described by Yau (1976). For this, take out a larger piece of skin covering three segments with the three ganglia, the connectives, and the roots intact.

Now observe the characteristic action of the motor neurons in the dissecting microscope. Some of the motor neurons are found on the ventral side of the ganglion, but the majority are on the dorsal surface. Use the paper by A.E. Stuart (1970) or Appendix D as a reference for the localization of the different motor neurons. In general, they are smaller than the sensory cells and perhaps a bit more difficult to impale. A few of them can be identified by inspection quite reliably; most of them will require a few trials before the right cell is found. The impulses of the motor neurons do not invade the cell body. The intracellular record from a motor neuron therefore looks quite unimpressive with impulses of only 2–10 mV in amplitude.

Start with the ventral side of the ganglion. The annulus erector (AE) cell is large and unambiguous. Impale the cell. Stop the impulse activity caused by the penetration by hyperpolarizing the cell body with a small (10^{-10} A) inward D.C. current. Inspect the skin in the dissecting microscope and watch the response when the current is reversed. Note the extent of the region innervated by the AE motor neuron.

The circular (CV) motor neuron is one of the larger cells just in front of the R cells. Determine its action and field of innervation in the same way.

Turn the ganglion with the dorsal aspect up. Examine the longitudinal (L) motor neuron. This is an important cell to know well since it will be used extensively later on.

Then look for some of the motor neurons that are less easily identified and evoke characteristic contractions — flattener motor neurons, oblique motor neurons, and dorsolateral and ventrolateral neurons.

Cut the posterior root of the ganglion and reexamine the peripheral fields of the AE, L, circular, and flattener motor neurons.

EXERCISE 6: STAINING SINGLE CELLS BY INTRACELLULAR INJECTION OF HORSERADISH PEROXIDASE

Certainly in the nervous system more than in any other organ, the shape of a cell is closely linked to its function. Whether studying biophysical aspects of membranes or the neuron as an integrative unit, the neurophysiologist is handicapped when ignorant of the morphology of the cell under study. Conversely, a knowledge of the cell's morphology can allow one to make informed guesses about the functioning cell. Over the past 15 years, the intracellular injection of dyes, such as Procion Yellow and of other markers such as Co^{++} into neurons has revolutionized neurophysiology. One marker that has worked especially well in the leech is the enzyme horseradish peroxidase (HRP). HRP diffuses in minutes or hours throughout the entire injected cell, catalyzes a reaction whose products are visible in both the light and electron microscope (permitting one to locate synapses), and does not seem to harm the injected cell in any measurable way. This exercise outlines the procedure for staining single cells by injecting them with HRP through the recording microelectrode. With practice, it will be possible to recognize the quality of the injection and become familiar with the features of some leech neurons and their synapses within the ganglion neuropil.

The following materials are needed for this exercise.

1. Recording chamber: Small plastic culture dish coated with a thin layer of transparent silicone rubber (Sylgard 184) and short minuten pins.

2. To be squirted into cells: Solution (previously filtered through 0.22 Millipore filter) of HRP (20 mg/ml; Sigma, Type VI) and Fast Green FCF, (2 mg/ml) in 0.2 M KCl to act as recording electrolyte.

3. Electrodes: Fiber-filled Pyrex-type capillary tubing (e.g., Haer 30-31-1) and electrode beveler fitted with 0.3 μm sandpaper (A.H. Thomas) (see Fig. 5).

4. Pressure Apparatus: Electrode holder (MPH-1, obtained from E.W. Wright, Guilford, Conn., with silver wire added for recording through electrode and side port tapped for pressure hose fitting); pressure valve, regulator, and gauge (Clippard, Cincinnati, Ohio) (see Fig. 6).

5. Fixative for tissue: Glutaraldehyde, 2%, in 0.1 M sodium phosphate buffer (pH 7.4, cold);

6. Reactants for stain: 5 mg diaminobenzidine tetrahydrochloride (DAB) dissolved in 10 ml of 0.1 M (pH 7.4) phosphate buffer; hydrogen peroxide, 1%.

7. For preparing whole mounts: Absolute ethanol (or methanol), xylene, mounting medium, slides, and coverslips.

Start by filling some microelectrodes with HRP solution; it sometimes takes an hour or so for the microelectrodes to fill completely. Thaw the vial of HRP solution (always kept in the freezer) in your hand. Pull six or so short-shank microelectrodes. They can be placed

FIGURE 5

▲ Microelectrode beveler.

on a strip of plasticine in a large Petri dish that is humidified by a bit of moist paper toweling. Next, using a long 30-g needle and small syringe, inject just a half centimeter or so of HRP solution into the shank of each microelectrode, taking care to avoid forming bubbles or wasting HRP (1 ml can fill more than 100 electrodes).

Put the microelectrode holder into the manipulator in place of the electrometer probe, which is moved to a nearby clamp. Then connect a lead from the probe to the pressure holder; this lead must be kept short to reduce picking up noise. The pressure tubing leading to the holder should be slack enough to permit the holder to be moved freely (but make certain that when maximal pressures are applied, the holder does not move, even by microscopic amounts). The valve and regulator assembly is fed by a suitable pressure source — usually the lab airline is sufficient (but be sure that the hose will not fly off).

Pin out the ganglion with cells to be injected in a Sylgard-coated dish containing a thin layer of leech Ringer's fluid. Then place the dish in the set-up on a thin piece of glass (e.g., a lantern slide glass) in place of the usual magnetic chamber. Be sure to tape down the glass; use some plasticine to secure the plastic dish and the ground electrode.

FIGURE 6

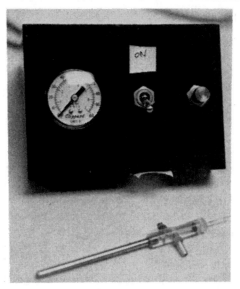

▲ Pressure injector.

Bevel a microelectrode. To do this, clamp the microelectrode in the beveler, lower the microelectrode under microscopic control so that just the tip is touching the spinning sandpaper for only about 10 sec (the time, of course, depends on speed, angle, surface, etc. — experience will tell what is best).

Thread the microelectrode onto the chlorided silver wire in the holder so that the wire dips into the HRP solution in the microelectrode. A twist of the end of the holder seals the microelectrode in place. Lower the microelectrode into the dish and measure the resistance. If it is more than 100 MΩ, it is too high and the microelectrode should be beveled some more; if it is under 50 MΩ, it is probably too low and it is unlikely that the cells will survive penetration. Between 60 MΩ and 100 MΩ is optimal for injecting leech neurons.

Now test whether the microelectrodes will inject. Pass current pulses of 0.5 nA through the microelectrode, balance out the microelectrode resistance with the bridge, and apply about 5–10 p.s.i. of pressure briefly (about 1 sec) to the microelectrode. While the pressure is being applied, the microelectrode resistance should drop 20% or more (bridge goes out of balance — which way?) and Fast Green might be seen flowing from the tip. If the resistance does not change or rises, the microelectrode is probably clogged. Try clearing the electrode with high pressure; if it does not clear, discard it. Once the microelectrode is working, make sure that the tip does not move when 10 or 15 p.s.i. pressure is applied. Movement usually can be eliminated by tightening the apparatus (especially the gasket screw that seals the electrode in place) or by changing the position of the pressure tubing that connects to the holder. Before penetrating the cell to be characterized try the electrode on a test cell (e.g., T cell). Large, beveled electrodes seem to penetrate best when they are lowered directly into the center of the cell with the fine drive of the manipulator. One person should monitor the oscilloscope screen, while the person penetrating the cell watches the electrode tip. If the T cell shows normal impulse pattern and the electrode is still working when removed, then the electrode is good for penetrating and injecting.

Once the cell sought is impaled, apply 5–10 p.s.i. for a few seconds, making sure that the electrode resistance drops. If it does not, the tip can sometimes be cleared by slight repositioning (backing off) or by a brief burst of high pressure. Now apply pressure longer. The cell should begin to turn green. The cell will get bigger; other signs of filling are an ever shorter time constant, and higher impulse rate (for "pagodas"), or wider and smaller impulses (sensory cells). Take care not to overinject; cells burst when their volume increases by more than 30–50%. If the cell is close to maximal volume and a dense stain is desired, let the cell recover for 5–10 minutes and reinject. Otherwise, remove the electrode and, if finished with the preparation, fill the dish, cover it, and allow the enzyme to diffuse for anywhere from 30 minutes to a few hours. Surprisingly, sometimes just an hour's diffusion is best, depending upon how extensive is the cell's arborization.

After diffusion of HRP throughout the cell, discard the Ringer's fluid and add about 2 ml fixative. After about 1/2 hour, wash 15 minutes in 0.1 M phosphate buffer. Incubate the tissue in DAB solution in a fume hood for about an hour (take care and use gloves as DAB is carcinogenic) and add a drop of 1% hydrogen peroxide, mixing with a Pasteur pipette. Follow the course of the reaction under a dissecting microscope. When the injected cell's processes become visible (at least in the roots and [or] connectives), but before the ganglion becomes brown, discard and rinse the DAB solution into a DAB waste beaker containing bleach in the hood and flood the preparation with phosphate buffer.

To dehydrate, replace buffer with 40% alcohol and then with three changes of 100% alcohol, leaving the tissue in the last for 4 to 5 minutes in a covered dish. (Keep the stock absolute alcohol well stoppered. Insufficient dehydration or hydrated alcohol will leave ganglia cloudy and sometimes covered with refractile droplets.)

Remove (or cut away) the pins holding the ganglion and rapidly transfer it to a glass or metal dish containing xylene. Now the preparation will become clear. Next place the ganglion (do not let it dry) into a drop of mounting medium on a microscope slide and coverslip.

To prepare the ganglion for electron microscopy, after rinsing away the DAB solution, fix the tissue for about 2 hours in 1% OsO_4, dehydrate in a more graded series of alcohols, and embed in Epon. The osmium treatment obscures the view through the whole ganglion, but when the ganglion is sectioned, the stained portions will be quite dark against the relatively unstained background. The DAB polymer is osmiophilic, and in the electron microscope is seen to fill densely the cytoplasm of the injected cell (but the stain does not cross intact membranes).

A camera lucida is used to draw the profile of the stained injected cell. This device, which is mounted between the objective lens and eyepiece of the microscope, contains a beam splitter that optically superimposes for the viewer both the image on a sheet of paper next to the microscope and the object in the microscope. Therefore, when the contours of the cell are traced with the tip of the simultaneously visible pen, the cell is "automatically" drawn. Use a stage micrometer to calibrate the dimensions of the cell on the paper. This magnification is somewhat adjustable and it is useful for comparing one cell with another always to use the same drawing tube magnification.

EXERCISE 7: LUCIFER YELLOW INJECTION

The fluorescent dye Lucifer Yellow (Stewart 1978) is also microinjected under pressure from electrodes filled with 3–10% Lucifer Yellow in 0.1 M $LiCl_2$. The dye can also be iontophoresed with hyperpolarizing pulses (5–10 nA) at a frequency of 0.5 Hz for 10–15 minutes. During a successful dye injection, the cell body turns bright yellow, as viewed in the darkfield optics used in electrophysiological experiments. After the dye has diffused through the cell for 15 minutes, the tissue is fixed for 1/2 hour in 4% paraformaldehyde in 0.1 M phosphate buffer, dehydrated, and put in a low-background fluorescence mounting medium. The ganglia can be viewed in methylsalicylate or immersion oil. However, to obtain a permanently mounted preparation, the tissue is transferred directly from 100% alcohol into a methacrylate embedding medium (Stewart 1978). After the embedding medium is polymerized with UV light, the Lucifer Yellow-labeled neuron can be viewed with fluorescein isothiocyanate (FITC) optics.

EXERCISE 8: RECORDING FROM AXONS IN INTACT ROOTS

To follow the natural history of the impulse, establish the function of a neuron, or study conduction block, it is useful to be able to record impulses of identified cells in intact roots supplying skin and muscle. There are several ways of doing this. All require that connective tissue should be cleaned off the root and that the electrodes be separated by extracellular fluid of high resistance.

1. The nerve can be placed on hook electrodes, and the Ringer fluid removed and replaced by mineral oil. This works well but inevitably results in oil covering one's fingers, the bath, the floor, etc.
2. A small hook electrode can be used to pull a loop of the nerve into a tube containing oil.
3. A suction electrode can be used to record *en passant*.
4. The simplest procedure is to use a bath containing two chambers separated by a partition with a gap through which the nerve runs. This gap is then sealed with a piece of photographic film cut to size and pressed into a groove containing Vaseline. A long stretch of nerve between ganglion and body wall works best; such a preparation is obtained by dissecting the dorsal branch of the posterior root that winds like a corkscrew between pockets of gut to innervate the dorsal body wall. First stroke the skin and observe impulses in the root. If all goes well and the seal around the nerve is good, the T cell impulses should be about 0.5–1.5 mV in amplitude.

Next identify the cell body of the T cell innervating dorsal skin (it is almost always the most lateral one in the ganglion) and impale it. Each pulse initiated in the cell body can be seen propagating out towards the skin and vice versa. By adjusting the interval between a stimulus to the cell body and a touch to the skin, the impulses can be made to collide, showing that they travel in the same axon. What is the approximate conduction velocity?

A harder task is to record the impulse of the lateral T cell that lies in the next ganglion and gives a small spike in the root. Look for P-cell and motor-cell impulses in the root. How do their sizes compare to those of T-cell impulses?

What conclusions can be drawn? The technique of identifying cells that are firing by the presence of their action potentials in various roots has been exploited with great success by Ort et al. (1974) in the swimming leech.

EXERCISE 9: SYNAPTIC POTENTIALS

A monosynaptic connection that is rewarding to study is the chemical synapse between the N sensory and the L motor cell. For this purpose pin the ganglion with its dorsal side uppermost. The N cell is situated laterally and can generally be seen at the end of the anterior packet near the root. The procedure is to impale the N cell before the L cell (why?). The excitatory potential with a good penetration in 7.5 mM Ca^{++} Ringer's fluid should be about 6 mV in amplitude or more. Measure:

1. The latency or delay between the N-cell impulse and the synaptic potential in the L cell.

2. The effect of hyperpolarizing the L cell on the amplitude of the synaptic potential.

3. Two-shock facilitation with different intervals.

4. The effect of brief trains of different frequencies. A cautionary note: This synapse becomes depressed quite rapidly, therefore do not stimulate with pairs or trains more than once every 15 seconds or so, preferably less often. Is there any electrical coupling between cell N and cell T?

Often, the lateral T cell will be found to occupy the place occupied by the N cell. If this occurs, see if there is a chemical EPSP in the L cell and test for electrical transmission. How does this synapse differ from that seen between R cells or between Leydig cells? Also examine the connection between the P and L cells. Sometimes, it is possible (difficult!) to impale three sensory cells in turn while recording from the L cell. Note that for these experiments sharp electrodes are essential.

In the same preparations, a good inhibitory synapse can be observed between the P cell and the "nut" (a cell of unknown function just in front of the AE cell). Measure the latency and the reversal potential using hyperpolarizing current pulses. In the T cells and HE cells, try to reverse the spontaneous IPSPs with current and also by injecting chloride into the cell by means of a 3 M KCl-filled electrode (about 10–50 MΩ).

A most useful technique is to record from cells while fluid flows past the ganglion. In this way, one can measure effects of changes in the concentration of Mg^{++}, K^+, Na^+, or Ca^{++} on membrane potentials and synaptic function. The flow works by gravity and suction. A good rate is about 1 ml/minute. Remember to check the fluid level in the reservoirs. From experience, it always tends to run out during a good experiment. Once the flow is going well, record the synaptic potential from the N cell to the L cell in fluid containing 7.5 mM Ca^{++}; then change to Ringer's fluid containing 20 mM Mg^{++} and 1.8 mM Ca^{++} for about 5 minutes; finally return to 7.5 mM Ca^{++} Ringer's fluid. Eventually, you should be able to work with flowing fluid routinely, but it requires some practice to control all the variables.

EXERCISE 10: TETRAETHYLAMMONIUM INJECTION

Tetraethylamonium (TEA) causes a decrease in K^+ conductance and a prolongation of the action potential. As a result, the longer action potential releases more transmitter at the nerve endings and this causes a larger synaptic potential. TEA injection therefore provides a good test for monosynaptic (compared to polysynaptic) connections. Use electrodes filled with 1–3 M TEA and low resistance (10–50 MΩ) or beveled to inject TEA into a T, P, and N cell. Once confident about the technique, observe the effect of TEA injection and the prolongation of the action potential in an N cell as well as the EPSP in the L cell. Also examine the connection between a P or N cell and an AE cell after injection of TEA into the presynaptic cells. The synaptic potential starts off as a minipotential, but see what happens.

REEMPHASIS OF CERTAIN PITFALLS AND HOW TO SOLVE THEM

1. Always be sure the light is optimal and the electrodes are sharp. Never use a microelectrode that blocks or whose resistance is too low.

2. If microelectrode after microelectrode appears to have an infinitely high resistance and not to be filled, check whether the bath is grounded.

3. Check that a 10 mV signal on the calibrator yields 1 cm deflection on the oscilloscope when the gain is 10 mV/cm. If not, or if resting and action potentials appear to have a small amplitude, check that the gain knob on the oscilloscope is in the calibrated position.

4. Excessive 60-cycle noise. Check lamps, transformers plugged into wall, bath electrode, ground loops, grounds to steel plate and microscope.

5. Do not use a ganglion that looks unhealthy or even slightly opaque, or granular. Check temperatures during dissection, heat filters, and

fluid levels, and change the fluid often. Only as a last resort blame the Ringer's fluid.

6. Before beginning an important experiment, always check the electrodes by impaling one or two easy cells (for example T or N cells). Remember that ringing a microelectrode may unblock it.

REFERENCES

Ort, C.A., W.B. Kristan, Jr., and G.S. Stent. 1974. Neuronal control of swimming in the medicinal leech. II. Identification of connections of motor neurons. *J. Comp. Physiol.* **94:**121–154.

Sawyer, R.T. 1972. *North American freshwater leeches, exclusive of the Piscicolidae*, 155 pp. University of Illinois Press, Urbana.

Stewart, W.W. 1978. Intracellular marking of neurons with highly fluorescent naphthalimide dye. *Cell* **14:**741–759.

Stuart, A.E. 1970. Physiological and morphological properties of motoneurones in the central nervous system of the leech. *J. Physiol.* **209:**627–646.

Thompson, W.J., and G.S. Stent. 1976. Neuronal control of the heart beat in the medicinal leech. I. Generation of the vascular constriction rhythm by heart motor neurons. *J. Comp. Physiol.* **111:**261–279.

Yau, K.-W. 1976. Receptive fields, geometry and conduction block of sensory neurones in the CNS of the leech. *J. Physiol.* **263:**513–538.

Appendix D

An Atlas of Neurons in the Leech, *Hirudo medicinalis*

The anatomy of the segmental ganglion is described in Chapter 4. Each ganglion is bilaterally symmetric and contains the monopolar cell bodies of nearly 200 pairs of homologous neurons. In addition, each ganglion contains a few unpaired cells, and many of those send axons into Faivre's nerve, the unpaired medial connective. The size and position of the cell bodies relative to the glial packet margins and other cell bodies in the ganglion is sufficiently stereotyped so that the cell body can often be recognized at sight, an aid in penetrating a previously identified cell with a microelectrode; otherwise it can be located as one of two or three possible candidates.

The maps of typical midbody segmental ganglia that are reproduced here (see Figs. 1 and 2) were devised as a system for naming all the cells of the ganglion based on position within the ganglion and without prejudice as to function (Ort et al. 1974). For example, cell bodies have been assigned numbers between 1 and 99 in the posterior lateral packet and between 101 and 199 in the anterior lateral packet. Bilateral homologs have been given the same number, and right and left members of any pair are distinguished (if necessary) by appending in parentheses the letters (R) and (L) to their numerical symbol. Several previously identified and intensively studied cells have been assigned letters rather than numbers.

FIGURE 1

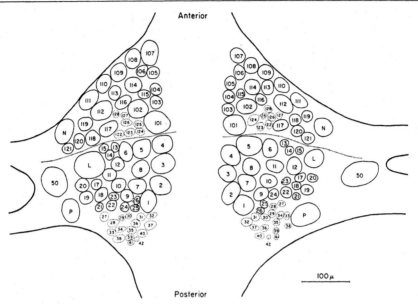

▲ Cell body map of the dorsal aspect of a midbody segmental ganglion of *H. medicinalis* showing the numbering system used to designate cell bodies. Cells with dashed outlines represent small cell bodies not always identifiable. The dashed line traversing the ganglion is the packet margin.

Although there is disagreement as to the aesthetics of such a numerical nomenclature, the present map has been widely used (see also Tables 1 and 2). Since 1974, when the maps were first drawn, accurate counts of cells in the ganglion have shown that complete maps will require the inclusion of several dozen more small cells to be distributed in all the glial packets. For this reason, the numbering system is necessarily provisional, and there remains at present a rather wide range in the degree of certainty to which a given number can be said to designate a particular neuron. Many of the numbered cells can be identified on the basis of their known pattern of branching, their impulse activity in the segmental nerves, and their known function. For these cells, the number designates a cell that is usually found to be one of two or three neighboring cells of similar size and electrical activity in a particular territory of the ganglion cortex. The cells drawn with dashed outlines represent neurons with very small cell bodies;

An Atlas of Neurons in the Leech, *Hirudo medicinalis*   **279**

FIGURE 2

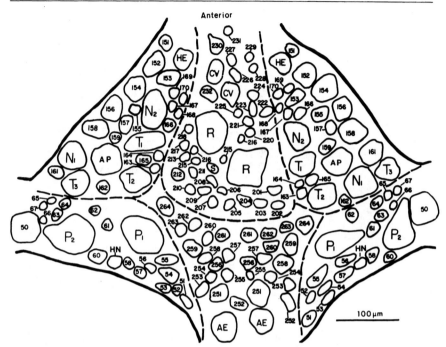

▲ Ventral aspect of a midbody segmental ganglion of *H. medicinalis*, showing the system used to designate cell bodies.

the numbers assigned to them are based entirely on approximate cell body location. Cells located at the edge of the ganglion are often visible both on dorsal and ventral aspects.

TABLE 1
Cells of leech segmental ganglion

Map number	Other designation	Description	Major axon projection [a,b]	References	Chapter number
Cells on the dorsal surface — Excitatory motor neurons					
3		excitor of dorsomedial longitudinal muscles	cr.	Ort et al. (1974)	7,8
4		excitor of ventromedial longitudinal muscles	ir.	Ort el al. (1974)	7,8
5	dl	excitor of dorsal longitudinal muscles	cr.	Stuart (1970); Ort et al. (1974)	5,7,8
7		excitor of dorsal longitudinal muscles	cr.	Ort et al. (1974)	7,8
8	vl	excitor of ventral longitudinal muscles	cr.	Stuart (1970); Ort et al. (1974)	5,7,8
11	v3	excitor of ventral circular muscles	cr.	Stuart (1970); Ort et al. (1974)	7,8
12	v1	excitor of ventrolateral circular muscles	cr.	Stuart (1970); Ort et al. (1974)	7,8
17		excitor of dorsolateral longitudinal muscles	cr.	B. Wallace (unpubl.)	
106	l	excitor of lateral longitudinal muscles	cr.	Stuart (1970); Ort et al. (1974)	5,7,8
107	d	excitor of dorsomedial longitudinal muscles	cr.	Stuart (1970); Ort et al. (1974)	5,7,8
108	v	excitor of ventromedial longitudinal muscles	cr.	Stuart (1970); Ort et al. (1974)	5,7
109	Fl	excitor of lateral dorsoventral muscles	cr.	Stuart (1970); Ort et al. (1974)	7
110	Ob	excitor of oblique muscles	cr.	Stuart (1970); Ort et al. (1974)	7,8
111	Ob	excitor of oblique muscles	cr.	Stuart (1970); Ort et al. (1974)	7
112	d	excitor of dorsal circular muscles	cr.	Stuart (1970); Ort et al. (1974)	7,8
117		excitor of medial dorsoventral muscles	ir.	B. Wallace (unpubl.)	
L		excitor of dorsal and ventral longitudinal muscles (large longitudinal motor neuron)	cr.	Stuart (1970); Ort et al. (1974)	5,6,7,8,10

Cells on the dorsal surface — Inhibitory motor neurons				
1	l-l-d	inhibitor of dorsal longitudinal muscles	cr.	Stuart (1970); Ort et al. (1974)	7,8
2	l-l-v	inhibitor of ventral longitudinal muscles	cr.	Stuart (1970); Ort et al. (1974)	7,8
101		inhibitor of dorsoventral muscles	cr.	Stuart (1970); Ort et al. (1970)	7,8
102	lfl	inhibitor of dorsal longitudinal muscles	cr.	Ort et al. (1970)	7,8
119		inhibitor of ventral longitudinal muscles	cr.	Sawada et al. (1976); Stent et al. (1978)	7
Cells on the dorsal surface — Interneurons				
27		interneuron in swimming oscillator	ic.	Stent et al. (1978)	7
28		interneuron in swimming oscillator	ic.	Stent et al. (1978)	7
33		interneuron in swimming oscillator	ic.	Stent et al. (1978)	7
(34)	Cl	coupling interneuron; electrically couples T cells with S cell and contralateral cells; on ventral or dorsal surface	none	Muller and Scott (1981)	6,A
123		interneuron in swimming oscillator	ic.	Stent et al. (1978)	7
Cells on the ventral surface — Excitatory motor neurons				
AE		excitor of subcutaneous muscle (annulus erector)	cr.	Stuart (1970)	5,6,8,10,A
CV	v2	excitor of ventrolateral circular muscles	cr.	Stuart (1970)	6
HE	avocado	excitor of the lateral heart tubes	ir.	Thompson and Stent (1976a)	5,6,7,8
Cells on the ventral surface — Sensory neurons				
T1 grape, T_l		mechanosensory, responds to touch on lateral third of skin	ir.ic(r).	Nicholls and Baylor (1968); Muller and McMahan (1976); Yau (1976b); Van Essen and Jansen (1977)	5,6,10
T2 grape, T_v		mechanosensory, responds to touch on ventral third of skin	ir.ic(r).	Nicholls and Baylor (1968); Muller and McMahan (1976); Yau (1976b); Van Essen and Jansen (1977)	5,6,10

TABLE 1 — *(Continued)*

Cells of leech segmental ganglion

Map number	Other designation	Description	Major axon projection [a,b]	References	Chapter number
T3	plum, T_d	mechanosensory, responds to touch on dorsal third of skin	ir.ic.(r).	Nicholls and Baylor (1968); Muller and McMahan (1976); Yau (1976b); Van Essen and Jansen (1977)	5,6,10
P1	potato, P_d	mechanosensory, responds to pressing on dorsal skin	ir.ic(r).	Nicholls and Baylor (1968); Muller and McMahan (1976)	5,6,8
P2	peach, P_v	mechanosensory, responds to pressing on ventral skin	ir.ic(r).	Nicholls and Baylor (1968); Muller and McMahan (1976)	5,6,8
N1	cuke, N_v	mechanosensory; responds to noxious stimuli on skin; innervates peripheral HO cells	ir.ic(r).	Nicholls and Baylor (1968); Muller and McMahan (1976)	5,6,8
N2	SF, N_d	mechanosensory; responds to noxious stimuli on skin and gut	ir.ic(r).	Nicholls and Baylor (1968); Muller and McMahan (1976)	5,6,8
Cells on the dorsal surface — Effectors					
R	Retzius, K_z	effects mucus release, swimming, relaxation of longitudinal muscles; contains 5-hydroxytryptamine (5-HT)	ir.ic(r).	Lent (1973); Mason and Leake (1978); Mason et al. (1979)	8
Cells on the ventral surface — Interneurons					
204	(unpaired)	interneuron initiates and maintains swimming	Fn.	Weeks and Kristan (1978)	7
205	(unpaired)	interneuron initiates and maintains swimming (found only in ganglion 9; probably homolgous with 104)	Fn.	Weeks (1980)	7
208	(unpaired)	interneuron in swimming oscillator	c.	Weeks (1980)	7

Cell	Description	Proj.	References	Fig.
S (unpaired)	interneuron contributes to shortening and swim initiation; "giant" axon is Rohde's fiber	Fn.	Frank et al. (1975); Magni and Pellegrino (1978a,b); Muller and Carbonetto (1979)	6,10
HN	inhibitory interneuron in heart oscillator (found in ganglia 1–7)	ic.	Thompson and Stent (1976b); Stent et al. (1979); Shafer and Calabrese (1981)	6,7
Unclassified or other neurons				
50 Leydig	unknown function; putative neurosecretory cell	ic(r).	Keyser (1979); Keyser et al. (1981); Stewart (1981)	8
153	unknown function; receives excitatory photosensory input	?	Kretz et al. (1976)	—
154	unknown function; receives excitatory photosensory input.	?	Kretz et al. (1976)	—
201	unknown function; inhibited by water vibration stimuli to sensilla	cc.	Friesen et al. (1981)	—
202	unknown function; excited by water vibration stimuli to sensilla	cc.	Friesen et al. (1981)	—
215	unknown function; receives excitatory photosensory input	?	Kretz et al. (1976)	—
216	unknown function; receives excitatory photosensory input	?	Kretz et al. (1976)	—
251 nut	unknown function; hyperpolarized by glutamate and P cell activity	ir.	Sargent (1975)	8
AP, A Pag.	unknown function; largest cell in anterior lateral glial packet ("anterior pagoda")	cr.	Sunderland (1980); Fuchs et al. (1981)	8

[a] Projections of axons from the ganglion are usually based on HRP injections, rather than electrical recording or Lucifer Yellow injections, which can give misleading results in the event of electrical or dye coupling to other neurons.

[b] Abbreviations: cr.=contralateral root(s); ir.=ipsilateral root(s); cc.=contralateral connective(s); c.=left or right or both connectives; ic.=ipsilateral connective(s); ic(r).=ipsilateral connective and root(s) of adjacent ganglion; Fn.=Faivre's nerve.

TABLE 2
Other leech neurons

Cell	Description	References	Chapter
Additional cells identified in segmental ganglia			
Neutral Red-staining cells	In each ganglion, there are three small cells that stain with Neutral Red dye and evidently contain 5-HT (see Chapter 8). All send processes into the lateral connectives but not into the roots of their own ganglion. One of the three (M) is unpaired and lies just anterior to the AE cell. The other two are on the dorsal (DL) and ventral (VL) surfaces in the posterior lateral packets and are positioned laterally. In the first three segmental ganglia, an additional paired cell with similar staining properties (cell E) is positioned just anteriorly to the R cells.	Stuart et al. (1974); Lent and Frazer (1977); Lent et al. (1979); Lent (1981)	8
Leu-enkephalin immunoreactive cell	An unpaired cell anterior to the AE cells found in *Haemopis marmorata* and believed to exist also in *Hirudo* stains with antibody to Leu-enkephalin.	Zipser (1980b)	8
Cells that bind monoclonal antibodies	Monoclonal antibodies have been prepared that can be used to stain particular cells not otherwise identified in *Haemopis* (e.g., the cell that binds Lan3-1).	Zipser and McKay, (1981)	B

The sex ganglia, numbers 5 and 6	In *Hirudo* and *Haemopis*, ganglia 5 and 6 contain nearly twice as many cells as the segmental ganglia in the midbody. Some cells innervating the sex organs have been identified as to function and by their connections.	Macagno (1980); Zipser (1979a,b, 1980) 5,B
The subesophageal ganglion	In the subesophageal ganglion, cells homologous with cells in the segmental ganglia have been found, including mechanosensory cells, R cells, and other neurons that stain with Neutral Red dye.	Stuart et al. (1974); Yau (1975, 1976a); Kleinhaus and Prichard (1974, 1975) 4,5
The supraesophageal ganglion	Some cells of the supraesophageal ganglion strongly scatter incident light and are considered to be neurosecretory. In *Macrobdella decora*, the morphology of some of these cells has been determined by staining them with injected HRP. Some cells of the supraesophageal ganglion of *Haemopis* stain specifically with particular monoclonal antibodies.	Orchard and Webb (1980); Zipser and McKay (1981) 4,B
Peripheral neurons	Sensory neurons responding to water waves and to light are found in the sensilla.	5,7
	In the body wall are large cells of unknown function (HO cells) that extend giant axons into the ganglion via the segmental nerves and are associated with endings of cell N2.	5
	A cell body is located at the first fork of the anterior segmental nerve it stains with Neutral Red dye and extends an axon into the ganglion that ramifies there and in the lateral connectives. The cell is believed to contain dopamine.	Stuart et al. (1974); Lent (1981) 5,8

REFERENCES

Frank, E., J.K.S. Jansen, and E. Rinvik. 1975. A multisomatic axon in the central nervous system of the leech. *J. Comp. Neurol.* **159**:1–13.

Friesen, W.O. 1981. Physiology of water motion detection in the medicinal leech. *J. Exp. Biol.* **92**:255–275.

Fuchs, P.A., J.G. Nicholls, and D.F. Ready. 1981. Membrane properties and selective connexions of identified leech neurones in culture. *J. Physiol.* (in press).

Keyser, K.T. 1979. Leydig cells within the c.n.s. of the leech. Ph.D. thesis, State University of New York, Stony Brook.

Keyser, K.T., B.M. Frazer, and C.M. Lent. 1981. Physiological and anatomical properties of Leydig cells in the segmental nervous system of the leech. *J. Comp. Physiol.* (in press).

Kleinhaus, A.L. and J.W. Prichard. 1974. Electrophysiological properties of the leech subesophageal ganglion. *Brain Res.* **72**:332–336.

──────── . 1975. Electrophysiological actions of convulsants and anticonvulsants on neurons of the leech subesophageal ganglion. *Comp. Biochem. Physiol.* **52C**:27–34.

Kretz, J.R., G.S. Stent, and W.B. Kristan, Jr. 1976. Photosensory input pathways in the medicinal leech. *J. Comp. Physiol.* **106**:1–37.

Lent, C.M. 1973. Retzius cells: Neuronal effectors controlling mucus release by the leech. *Science* **179**:693–696.

──────── . 1981. Morphology of neurons containing monoamines within leech segmental ganglia. *J. Exp. Zool.* **216**:311–316.

Lent, C.M. and B.M. Frazer. 1977. Connectivity of the monoamine-containing neurones in central nervous system of leech. *Nature* **266**:844–847.

Lent, C.M., J. Ono., K.T. Keyser, and H.J. Karten. 1979. Identification of serotonin within vital stained neurons from leech ganglia. *J. Neurochem.* **32**:1551–1564.

Macagno. E.R. 1980. Number and distribution of neurons in leech segmental ganglia. *J. Comp. Neurol.* **190**:283–302.

Magni, F. and M. Pellegrino, 1978a. Patterns of activity and the effects of activation of the fast conducting system on the behavior of unrestrained leeches. *J. Exp. Biol.* **76**:123–135.

──────── . 1978b. Neural mechanisms underlying the segmental and generalized cord shortening reflexes in the leech. *J. Comp. Physiol.* **124**:339–351.

Mason, A. and L.D. Leake. 1978. Morphology of leech Retzius cells demonstrated by intracellular injection of horseradish peroxidase. *Comp. Biochem. Physiol.* **61A**:213–216.

Mason, A., A.J. Sunderland, and L.D. Leake. 1979. Effects of leech Retzius cells on body wall muscles. *Comp. Biochem. Physiol.* **63C**.359–361.

Muller, K.J. and S.T. Carbonetto. 1979. The morphological and physiological properties of a regenerating synapse in the C.N.S. of the leech. *J. Comp. Neurol.* **185**:485–516.

Muller, K.J. and U.J. McMahan. 1976. The shapes of sensory and motor neurones and the distribution of their synapses in ganglia of the leech: a study using intracellular injection of horseradish peroxidase. *Proc. R. Soc. Lond. B* **194**:481–499.

Muller, K.J. and S.A. Scott. 1981. Transmission at a "direct" electrical connexion mediated by an interneurone in the leech. *J. Physiol.* **311**:565–583.

Nicholls, J.G. and D.A. Baylor, 1968. Specific modalities and receptive fields of sensory neurons in the CNS of the leech. *J. Neurophysiol.* **31**:740–756.

Orchard, I. and R.A. Webb. 1980. The projections of neurosecretory cells in the brain of the North-American medicinal leech, *Macrobdella decora*, using intracellular injection of horseradish peroxidase. *J. Neurobiol.* **11**:229–242.

Ort, C.A., W.B. Kristan, and G.S. Stent. 1974. Neuronal control of swimming in the medicinal leech. II. Identification and connections of motor neurons. *J. Comp. Physiol.* **94**:121–154.

Sargent, P.B. 1975. "Transmitters in the leech central nervous systems: Analysis of sensory and motor cells." Ph.D. thesis, Harvard University, Cambridge, Massachusetts.

Sawada, M., J.M. Wilkinson, D.J. McAdoo, and R.E. Coggeshall. 1976. The identification of two inhibitory cells in each segmental ganglion of the leech and studies on the ionic mechanism of the inhibitory junctional potentials produced by these cells. *J. Neurobiol.* **7**:435–445.

Shafer, M.R. and R.L. Calabrese. 1981. Similarities and differences in the structure of segmentally homologous neurons that control the hearts in the leech, *Hirudo medicinalis*. *Cell Tissue Res.* **214**:137–153.

Stent, G.S., W.J. Thompson, and R.L. Calabrese. 1979. Neural control of heartbeat in the leech and in some other invertebrates. *Physiol. Revs.* **59**:101–136.

Stent, G.S., W.B. Kristan, Jr., W.O. Friesen, C.A. Ort, M. Poon, R.L. Calabrese. 1978. Neuronal generation of the leech swimming movement. *Science* **200**:1348–1357.

Stewart W.W. 1981. Lucifer dyes — highly fluorescent dyes for biological tracing. *Nature* **292**:17–21.

Stuart, A.E. 1970. Physiological and morphological properties of motoneurones in the central nervous system of the leech. *J. Physiol.* **209**:627–646.

Stuart, A.E., A.J. Hudspeth, and Z.W. Hall. 1974. Vital staining of specific monoamine-containing cells in the leech nervous system. *Cell Tissue Res.* **153**:55–61.

Sunderland, A.J. 1980. A hitherto undocumented pair of neurons in the segmental ganglion of the leech which receive synaptic input from mechanosensory cells. *Comp. Biochem. Physiol.* **67A**:299–302.

Thompson, W.J. and G.S. Stent. 1976a. Neuronal control of heartbeat in the medicinal leech I. Generation of the vascular constriction rhythm by heart motor neurons. *J. Comp. Physiol.* **111**:261–279.

———. 1976b. Neuronal control of heartbeat in the medicinal leech II. Intersegmental coordination of heart motor neuron activity by heart interneurons. *J. Comp. Physiol.* **111**:281–307.

Van Essen, D.C. and J.K.S. Jansen. 1977. The specificity of re-innervation by identified sensory and motor neurons in the leech. *J. Comp. Neurol.* **171**:433–454.

Weeks, J.C. 1980. "The roles of identified interneurons in initiating and generating the swimming motor pattern of leeches." Ph.D. thesis, University of California, La Jolla.

Weeks, J.C. and W.B. Kristan Jr. 1978. Initiation, maintenance and modulation of swimming in the medicinal leech by the activity of a single neuron. *J. Exp. Biol.* **77**:71–88.

Willard, A.L. 1981. Effects of serotonin on the generation of the motor program for swimming by the medicinal leech. *J. Neurosci.* (in press).

Yau, K.-W. 1975. "Receptive fields, geometry and conduction block of sensory cells in the leech central nervous system." Ph.D. thesis, Harvard University, Cambridge, Massachusetts.

Yau, K.-W. 1976a. Physiological properties and receptive fields of mechanosensory neurones in the head ganglion of the leech: Comparison with homologous cells in segmental ganglia. *J. Physiol.* **263**:489–512.

———. 1976b. Receptive fields, geometry and conduction block of sensory neurones in the CNS of the leech. *J. Physiol.* **263**:513–538.

Zipser, B. 1979a. Identifiable neurons controlling penile eversion in the leech. *J. Neurophysiol.* **42**:455–464.

———. 1979b. Voltage-modulated membrane resistance in coupled leech neurons. *J. Neurophysiol.* **42**:465–475.

———. 1980a. Horseradish peroxidase nerve backfilling in leech. *Brain Res.* **182**:441–445.

———. 1980b. Identification of specific leech neurones immunoreactive to enkephalin. *Nature* **283**:857–858.

Zipser, B. and R. McKay. 1981. Monoclonal antibodies distinguish identifiable neurones in the leech. *Nature* **289**:549–554.

Bibliography

Adam, A. 1968. Simulation of rhythmic nervous activities. II. Mathematical models for the function of networks with cyclic inhibition. *Kybernetik* **5**:103–109. (7)

Albucassis. c 1013. *On surgery and instruments.* (Translated by M.S. Spink and G.L. Lewis, 1973.) University of California Press, Berkeley. (3)

Alving, B.O. 1968. Spontaneous activity in isolated somata of *Aplysia* pacemaker neurons. *J. Gen. Physiol.* **45**:29–45. (7)

Anderson, D.T. 1973. *Embryology and phylogeny in annelids and arthropods.* Pergamon Press, Oxford. (2)

Andrew, W. 1965. *Comparative haematology.* Grune and Stratton, New York. (2)

Armett-Kibel, C., I.A. Meinertzhagen, and J.E. Dowling. 1977. Cellular and synaptic organization in the lamina of the dragon-fly, *Sympetrum rubicundulum*. *Proc. R. Soc. Lond. B* **196**:385–413. (6)

Ascher, P. 1972. Inhibitory and excitatory effects of dopamine on *Aplysia* neurones. *J. Physiol.* **225**:173–209. (6)

Atwood, H.L. and G.D. Bittner. 1971. Matching of excitatory and inhibitory inputs to crustacean muscle fibers. *J. Neurophysiol.* **34**:157–170. (6)

Autrum, H. 1939. Literatur uber Hirudineen. In *Klassen* and *Ordnungen des Tierreichs*, (ed H.G. Bronns), 4, III, 4 Hirudineen. (3)

The numbers or letters in parentheses indicate the chapters in which a particular reference is cited.

Bacq, Z.M. and G. Coppée. 1937. Action de l'ésérine sur la préparation neuromusculaire du siponcle et de la sangsue. *C. R. Soc. Biol.* **124:**1244–1247. (8)

Bagdy, D., E. Barabas, L. Graf, T.E. Petersen, and S. Magnusson. 1976. Hirudin. *Methods Enzymol.* **45:**669–678. (2)

Batelle, B.-A. and E.A. Kravitz. 1978. Targets of octopamine action in the lobster: Cyclic nucleotide changes and physiological effects in haemolymph, heart and exoskeletal muscle. *J. Pharmacol. Exp. Ther.* **205:**438–448. (8)

Baylor, D.A. and J.G. Nicholls. 1969. After-effects of nerve impulses on signalling in the central nervous system of the leech. *J. Physiol.* **203:**571–589. (6)

―――― . 1969. Chemical and electrical synaptic connexions between cutaneous mechanoreceptor neurones in the central nervous system of the leech. *J. Physiol.* **203:**591–609. (5,6)

―――― . 1969. Changes in extracellular potassium concentration produced by neuronal activity in the central nervous system of the leech. *J. Physiol.* **203:**555–569. (4)

―――― . 1969. Long-lasting hyperpolarization after activity of neurons in leech central nervous system. *Science* **162:**279–281. (5,6)

―――― . 1971. Patterns of regeneration between individual nerve cells in the central nervous system of the leech. *Nature* **232:**268–270. (10)

Beeton, I. 1861. *The book of household management.* (Facsimile, 1969.) Cape, London. (3)

Bennett, M.V.L. 1973. Function of electrotonic junctions in embryonic and adult tissues. *Fed. Proc.* **32:**65–75. (6)

Berry, M.S. and V.W. Pentreath. 1976. Properties of a symmetric pair of serotonin-containing neurones in the cerebral ganglia of *Planorbis*. *J. Exp. Biol.* **65:**361–380. (8)

Bianchi, S. 1967. On the different types of fluorescent neurons in the leech (*Hirudo medicinalis*). *Atti. Soc. Peloritana. Sci. Fis. Mat. Nat.* **13:**39–47. (8)

―――― . 1974. The histochemistry of the biogenic amines in the central nervous system of *Hirudo medicinalis*. *Gen. Comp. Endocrinol.* **22:**245–249. (8)

Biedl, A. 1910. Ueber das Adrenalgewebe bei Wirbellosen. *Int. Congr. Zool. Graz.* **8:**503–505. (8)

Birks, R.I. and F.C. MacIntosh. 1961. Acetylcholine metabolism of a sympathetic ganglion. *Can. J. Biochem. Physiol.* **39:**787–827. (8)

Blackshaw, S. 1981. Morphology and distribution of touch cell terminals in the skin of the leech. *J. Physiol.* (in press). (5,8)

Blair, S.S. 1981. Interactions between mesoderm and ectoderm in segment formation in the embryo of a glossiphoniid leech. *Dev. Biol.* (in press). (9)

Blair, S.S. and D.A. Weisblat. 1982. Ectodermal interaction during neurogenesis in the glossiphoniid leech, *Helobdella triserialis*. *Dev. Biol.* (in press). (9)

Blankenship, J.E., H. Wachtel, and E.R. Kandel. 1971. Ionic mechanisms of excitatory, inhibitory, and dual synaptic actions mediated by an identified interneuron in abdominal ganglion of *Aplysia*. *J. Neurophysiol.* **34:**76–92. (8)

Boaistuau, P. 1560. *Histoires prodigieuses.* Wellcome M.S., London, WHMM 6573. (3)

Bon, S., M. Vigny, and J. Massoulié. 1979. Asymmetric and globular forms of acetylcholinesterase in mammals and birds. *Proc. Natl. Acad. Sci.* **76**:2546–2550. (8)

Boroffka, I. and R. Hamp. 1969. Topographie des Kreislaufsystems und Zirkulation bei *Hirudo medicinalis*. *Z. Morphol. Tiere.* **64**:59–76. (7)

Bowling, D., J. Nicholls, and I. Parnas. 1978. Destruction of a single cell in the C.N.S. of the leech as a means of analyzing its connexions and functional role. *J. Physiol.* **282**:169–180. (6,A)

Bradbury, S. 1959. The botryoidal and vasofibrous tissue of the leech, *Hirudo medicinalis*. *Q. J. Microsc. Sci.* **100**:483–498. (4)

Bradley, G.W., C. von Euler, I. Marttila, and B. Roos. 1975. A model of the central and reflex inhibition of inspiration in the cat. *Biol. Cybern.* **19**:105–116. (7)

Brimijoin, S., K. Skau, and M.J. Wiermaa. 1978. On the origin and fate of external acetylcholinesterase in peripheral nerve. *J. Physiol.* **285**:143–158. (8)

Brockbank, W. 1954. *Ancient therapeutic arts.* Heinemann, London. (3)

Brown, T.G. 1911. The intrinsic factors in the act of progression in the mammal. *Proc. R. Soc. Lond. B* **84**:308–319. (7)

Brownstein, M.J., J.M. Saavedra, J. Axelrod, G.H. Zeman, and D.O. Carpenter. 1974. Coexistence of several putative neurotransmitters in single identified neurons of *Aplysia*. *Proc. Natl. Acad. Sci.* **71**:4662–4665. (8)

Buchan, W. 1788. *Domestic medicine*, 10th Edition. A. Strahan, London. (3)

Budzynski, A.Z., S.A. Olexa, B.S. Brizuela, R.T. Sawyer, and G.S. Stent. 1981. Anticoagulant and fibrinolytic properties of salivary proteins from the leech *Haementeria ghilianii*. *Proc. Soc. Exp. Biol. Med.* (in press). (2)

Bullock, T.H. 1961. The origins of patterned nervous discharge. *Behavior* **17**:48–59. (7)

Büsing, K.H. 1951. *Pseudomonas hirudinis* ein baktrieller Darmsymbiont des Blutegels (*Hirudo medicinalis*). *Zentralbl. Bakteriol. I Orig.* **157**:478–484. (4)

Büsing, K.H., W. Döll, and K. Freytag. 1953. Die Bakterienflora del medizinischen Blutegel. *Arch. Mikrobiol.* **19**:52–86. (4)

Calabrese, R.L. 1977. The neural control of alternate heartbeat coordination states in the leech, *Hirudo medicinalis*. *J. Comp. Physiol.* **122**:111–143. (7)

―――――. 1977. Regeneration of an intersegmental interneuron in the leech. *Neurosci. Abstr.* **3**:296. (10)

―――――. 1979. Neural generation of the peristaltic and nonperistaltic heartbeat coordination modes of the leech, *Hirudo medicinalis*. *Am. Zool.* **19**:87–102. (7)

―――――. 1979. The roles of endogenous membrane properties and synaptic interaction in generating the heartbeat rhythm of the leech, *Hirudo medicinalis*. *J. Exp. Biol.* **82**:163–176. (7)

―――――. 1980. Control of impulse-initiation sites in a leech interneuron. *J. Neurophysiol.* **44**:878–896. (7)

Cammelli, E., A.M. DeBellis, and A. Nistri. 1974. Distribution of acetylcholine and of acetylcholinesterase activity in the nervous tissue of the frog and of the leech. *J. Physiol.* **242**:88–90. (8)

Carbonetto, S. and K.J. Muller. 1977. A regenerating neurone in the leech can form an electrical synapse on its severed axon segment. *Nature* **267**:450–452. (6,10)

Clowes, W. 1596. *A profitable and necessary book of observations.* Scholars' Facsimiles and Reprints, New York. (3)

Cockayne, O. 1864. *Leechdoms, wortcunning and starcraft,* vol. 1–3. Revised ed. (Introduction by C. Singer, 1961.) Holland Press, London. (3)

Coggeshall, R.E. 1972. Autoradiographic and chemical localization of 5-hydroxytryptamine in identified neurons in the leech. *Anat. Record.* **172**:489–498. (8)

———. 1974. Gap junctions between identified glial cells in the leech. *J. Neurobiol.* **5**:463–467. (6)

Coggeshall, R.E. and D.W. Fawcett. 1964. The fine structure of the central nervous system of the leech, *Hirudo medicinalis. J. Neurophysiol.* **27**:229–289. (4,6,8)

Coggeshall, R.E., S.A. Dewhurst, D. Weinreich, and R.E. McCaman. 1972. Aromatic acid decarboxylase and choline acetylase activities in a single identified 5-HT containing cell of the leech. *J. Neurobiol.* **3**:259–265. (8)

Cottrell, G.A. 1974. Serotonin and free amino acid analysis of ganglia and isolated neurones of *Aplysia dactylomelia. J. Neurochem.* **22**:557–559. (8)

———. 1976. Does the giant cerebral neurone of *Helix* release two transmitters: ACh and serotonin? *J. Physiol.* **259**:44–45P. (8)

Couteaux, R. and M. Pecot-Dechavassine. 1970. Vesicules synaptiques et poches au niveau des 'zones actives' de la jonction neuromusculaire. *C. R. Acad. Sci. D* **271**:2346–2349. (6)

Daniels, B.A. 1975. "On the biology of *Actinobdella inequiannulata* (Hirudinea, Glossiphoniidae) parasitic on *Catostomus* spp. in Algonquin Park." Masters thesis, University of Toronto, Ontario. (2)

Daniels, B.A. and R.T. Sawyer. 1975. The biology of the leech *Myzobdella lugubris* infesting blue crabs and catfish. *Biol. Bull.* **148**:193–198. (2)

Davies, R.W., F.J. Wrona, and F. Linton. 1979. A serological study of prey selection by *Helobdella stagnalis* (Hirudinoidea). *J. Animal Ecol.* **48**:181–194. (2)

Dawson, W.R. 1934. *A leechbook on collection of medical recipes of the 15th century.* London. (3)

del Castillo, J. and B. Katz. 1954. Statistical factors involved in neuromuscular facilitation and depression. *J. Physiol.* **124**:574–585. (6)

Delcomyn, F. 1980. Neural basis of rhythmic behavior in animals. *Science* **210**:492–498. (7)

Derganc, M. and F. Zdravic. 1960. Venous congestion of flaps treated by application of leeches. *Bri. J. Plast. Surg.* **13**:187–192. (4)

DeRiemer, S.A. and E.R. Macagno. 1981. Light microscopic analysis of contacts between pairs of identified leech neurons with combined use of horseradish peroxidase and Lucifer Yellow. *J. Neurosci.* **1**:650–657. (6)

Derosa, Y.S. and W.O. Friesen. 1981. Morphology of leech sensilla: Observations with the scanning electron microscope. *Biol. Bull.* **160**:(in press). (4)

Deschènes, M. and M.V.L. Bennett. 1974. A qualification to the use of TEA as

a tracer for monosynaptic pathways. *Brain Res.* **77**:169–172. (6)
Dodge, F.A., B.W. Knight, and J. Toyoda. 1968. Voltage noise in *Limulus* visual cells. *Science* **160**:88–90. (6)
Dowling, J.E. 1970. Organization of vertebrate retinas. *Invest. Ophthalmol.* **9**:655–680. (6)
Dreyer, F., K. Peper, K. Akert, C. Sandri, and H. Moor. 1973. Ultrastructure of the "active zone" in the frog neuromuscular junction. *Brain Res.* **62**:373–380. (6)
Dudel, J. and S.W. Kuffler. 1961. Presynaptic inhibition at the crayfish neuromuscular junction. *J. Physiol.* **155**:543–562. (6)
Dudel, J., Y. Grossman, and I. Parnas. 1980. Synaptic transmission in crustacean muscle; effects of elimination of the inhibitor fiber on excitatory transmission. In *Amino acid transmitters* (ed. P. Mandel and F.V. Defeudis). Raven Press, New York. (A)
Eckert, R. 1963. Electrical interaction of paired ganglion cells in the leech. *J. Gen. Physiol.* **46**:573–587. (6)
Ehinger, B., B. Falck, and H.E. Myhrberg. 1968. Biogenic monoamines in *Hirudo medicinalis*. *Histochemie* **15**:140–149. (8)
Elliott, E.J. and K.J. Muller. 1981. Long-term survival of glial segments during nerve regeneration in the leech. *Brain Res.* **218**:99–114. (10)
Evans, P.D., E.A. Kravitz, B.R. Talamo, and B.G. Wallace. 1976. The association of octopamine with specific neurones along lobster nerve trunks. *J. Physiol.* **262**:51–70. (8)
Falck, B., N.-A. Hillarp, G. Thieme, and A. Torp. 1962. Fluorescence of catecholamines and related compounds condensed with formaldehyde. *J. Histochem. Cytochem.* **10**:348–354. (8)
Fambrough, D.M. 1979. Control of acetylcholine receptors in skeletal muscle. *Physiol. Rev.* **59**:165–227. (8)
Fernández, J.H. 1975. Structure and arrangement of synapses in the C.N.S. of the leech under normal and experimental conditions. *Neurosci. Abstr.* **1**:1023. (6)
―――. 1980. Embryonic development of the glossiphoniid leech *Theromyzon rude*: Characterization of developmental stages. *Dev. Biol.* **76**:245–262. (9)
Fernández, J.H. and M.S.G. Fernández. 1974. Morphological evidence for an experimentally induced synaptic field. *Nature* **251**:428–430. (10)
Fernández, J. and G.S. Stent. 1980. Embryonic development of the glossiphoniid leech *Theromyzon rude*: Structure and development of the germinal bands. *Dev. Biol.* **78**:407–434. (9)
Fett, M.J. 1978. Quantitative mapping of cutaneous receptive field, in normal and operated leeches, *Linnobdella*. *J. Exp. Biol.* **76**:167–179. (A)
Flacke, W. and T.S. Yeoh. 1968. The action of some cholinergic antagonists and anticholinesterase agents on the dorsal muscle of the leech. *Br. J. Pharmacol. Chemother.* **33**:145–153. (8)
Frank, E. 1973. Matching of facilitation at the neuromuscular junction of the lobster: A possible case for influence of muscle on nerve. *J. Physiol.* **233**:635–658. (6)

Frank, E., J.K.S. Jansen, and E. Rinvik. 1975. A multisomatic axon in the central nervous system of the leech. *J. Comp. Neurol.* **159**:1–13. (6,D)

Frazier, W.T.E., E.R. Kandel, I. Kupfermann, R. Waziri, and R.E. Coggeshall. 1967. Morphological and functional properties of identified neurons in the abdominal ganglion of *Aplysia californica*. *J. Neurophysiol.* **30**:1288–1351. (7)

Friesen, W.O. 1981. Physiology of water motor detection in the medicinal leech. *J. Exp. Biol.* **92**:255–275. (7,D)

Friesen, W.O. and R. D. Dedwylder. 1978. Detection of low amplitude water movement: A new sensory modality in the medicinal leech. *Neurosci. Abstr.* **4**:380. (2)

Friesen, W.O. and G.S. Stent. 1977. Generation of a locomotory rhythm by a neural network with recurrent cyclic inhibition. *Biol. Cybern.* **28**:27–40. (7)

———. 1978. Neuronal circuits for generating rhythmic movements. *Ann. Rev. Biophys. Biochem.* **7**:37–61. (7)

Friesen, W.O., M. Poon, and G.S. Stent. 1976. An oscillatory neuronal circuit generating a locomotory rhythm. *Proc. Natl. Acad. Sci.* **73**:3734–3738. (6)

———. 1978. Neuronal control of swimming in the medicinal leech. IV. Identification of a network of oscillatory interneurones. *J. Exp. Biol.* **75**:25–43. (6,7)

Fuchs, P.A., J.G. Nicholls, and D.F. Ready. 1981. Membrane properties and selective connexions of identified leech neurones in culture. *J. Physiol.* **316**:203–223. (5,8,10,D)

Führner, J. 1917. Ein Vorlesungsversuch zur Demonstration der erregbarkeitssteigernden Wirkung des Physostigmins. *Arch. Exp. Pathol. Pharmokol.* **82**:81–85. (8)

Gardner, D. and E.R. Kandel. 1977. Physiological and kinetic properties of cholinergic receptors activated by multiaction interneurons in buccal ganglia of *Aplysia*. *J. Neurophysiol.* **40**:333–348. (6)

Gardner-Medwin, A.R., J.K.S. Jansen, and T. Taxt. 1973. The "giant" axon of the leech. *Acta Physiol. Scand.* **87**:30A–31A. (6)

Gaskell, J.F. 1914. The chromaffine system of annelids and the relation of this system to the contractile vascular system in the leech, *Hirudo medicinalis*. *Philos. Trans. R. Soc. Lond. B* **205**:153–212. (8)

Geffen, L.B. and B.G. Livett. 1971. Synaptic vesicles in sympathetic neurons. *Physiol. Rev.* **51**:98–157. (6)

Gerschenfeld, H.M. 1973. Chemical transmission in invertebrate central nervous systems and neuromuscular junctions. *Physiol. Rev.* **53**:1–119. (6,8)

Gerschenfeld, H.M. and D. Paupardin-Tritsch. 1974. Ionic mechanisms and receptor properties underlying the responses of molluscan neurones to 5-hydroxytryptamine. *J. Physiol.* **243**:427–456. (8)

Giller, E. and J.H. Schwartz. 1971. Acetylcholinesterase in identified neurons of abdominal ganglion of *Aplysia californica*. *J. Neurophysiol.* **34**:108–115. (8)

———. 1971. Choline acetyltransferase in identified neurons of abdominal ganglion of *Aplysia californica*. *J. Neurophysiol.* **34**:93–107. (8)

Goodman, C.S., M. O'Shea, R. McCaman, and N.C. Spitzer. 1979. Embryonic development of identified neurons: Temporal pattern of morphological and biochemical differentiation. *Science* **204**:1219–1222. (8)

Gorman, A.L.F., A. Hermann, and M.V. Thomas. 1980. The neuronal pacemaker cycle. In *Molluscan nerve cells: From biophysics to behavior* (ed. J. Koester and H.J. Byrne), pp. 169–180. Cold Spring Harbor Laboratory, Cold Spring Harbor, New York. (7)

Gratiolet, P. 1862. Recherches sur l'organisation du système vasculaire dans la sangsue medicinale et l'aulastome vorace. *Ann. Sci. Nat. Zool.* **17**:174–225. (4)

Gray, J. 1968. *Animal locomotion*. Weidenfels and Nicholson, London. (7)

Gray, J., H.W. Lissman, and R.J. Pumphrey. 1938. The mechanism of locomotion in the leech (*Hirudo medicinalis*). *J. Exp. Biol.* **15**:408–430. (2,7)

Grinvald, A., L.B. Cohen, S. Lesher, and M.B. Boyle. 1981. Simultaneous optical monitoring of activity of many neurons in invertebrate ganglia using a 124-element photodiode array. *J. Neurophysiol.* **45**:829–840. (5)

Gruner, O.C. 1930. *A treatise on the Canon of Avicenna, incorporating a translation of the first book*. Luzac, London. (3)

Hagiwara, S. and H. Morita. 1962. Electrotonic transmission between two nerve cells in the leech ganglion. *J. Neurophysiol.* **25**:721–731. (6)

Hall, Z.W. 1973. Multiple forms of acetylcholinesterase and their distribution in endplate and non-endplate regions of rat diaphragm muscle. *J. Neurobiol.* **4**:343–361. (8)

Hall, Z.W., M.D. Bownds, and E.A. Kravitz. 1970. The metabolism of gamma aminobutyric acid in the lobster nervous system. Enzymes in single excitatory and inhibitory axons. *J. Cell Biol.* **46**:290–299. (8)

Hanley, M.R., G.A. Cottrell, P.C. Emson, and F. Fonnum. 1974. Enzymatic synthesis of acetylcholine by a serotonin-containing neurone from *Helix*. *Nature* **251**:631–633. (8)

Harant, H. and P.-P. Grassé. 1959. Classe des annélidees achètes ou hirudinées ou sangsues. In *Traité de zoologie*. (ed. P.-P. Grassé), vol. 5, pp. 471–595. Masson, Paris. (4)

Harmon, L.D. and E.R. Lewis. 1966. Neural modeling. *Physiol. Rev.* **46**:513–591. (7)

Hartnett, J.C. 1972. The care and use of medicinal leeches in 19th century pharmacy and therapeutics. *Pharm. Hist.* **14**:127–138. (3)

Harvey, W. 1628. *De Moku Cordis*. G. Fitzeri, Frankfurt. (4)

Haycraft, J.B. 1884. On the action of a secretion obtained from the medicinal leech on the coagulation of blood. *Proc. Roy. Soc.* **36**:478–487. (3,4)

Henderson, L. 1981. Serotonergic transmission between isolated leech neurones in culture. *Neurosci. Abstr.* (in press). (8)

Herter, K. 1932. Hirudinea, Egel. In *Biologie der Tiere Deutschlands* (ed. P. Schulze), vol. 12b, pp. 1–158. Berlin. (2)

Heunius, J. 1652. *De hirudinum usu et efficacia in medicina tractatus*. J. Jegeri, Gryphiswaldiae. (3)

Heuser, J.E., T.S. Reese, and D.M.D. Landis. 1974. Functional changes in frog neuromuscular junctions studied with freeze-fracture. *J. Neurocytol.* **3**:109–131. (6)

Hildebrand, J.G., J.G. Townsel, and E.A. Kravitz. 1974. Distribution of acetylcholine, choline, choline acetyltransferase and acetylcholinesterase in regions and single identified axons of the lobster nervous system. *J. Neurochem.* **23**:951–963. (8)

Hildebrand, J.G., D.L. Barker, E. Herbert, and E.A. Kravitz. 1971. Screening for neurotransmitters: A rapid radiochemical procedure. *J. Neurobiol.* **2**:231–246. (8)

Hökfelt, T., O. Johansson, Å. Ljungdahl, J.M. Lundberg, and M. Schultzberg. 1980. Peptidergic neurones. *Nature* **284**:515–521. (8)

Horn, G. 1798. *An entire new treatise on leeches wherein the singular and valuable reptile is most clearly set forth.* London. (3)

Hoy, R.R., G.D. Bittner, and D. Kennedy. 1967. Regeneration of crustacean motoneurons: Evidence of axonal fusion. *Science* **156**:251–252. (A)

Hubbard, J.I. 1974. Neuromuscular transmission — Presynaptic factors. In *The peripheral nervous system* (ed. J. I. Hubbard), pp. 151–180. Plenum Press, New York. (6)

Huber, J. 1891. *Nicander — Theriaca.* apud Guil. Morelium. *Dtsch Arch. Klin Med.* **67**:522. (3)

Ito, T. 1936. Zytologische Untersuchungen ueber die Ganglienzellen des Japanischen medizinischen Blutegels, *Hirudo nipponica*, mit besonderer Berucksichtigung auf die "dunkle Ganglienzelle." *Okajimas Folia Anat. Jpn.* **14**:111–169. (8)

Jansen, J.K.S. and J.G. Nicholls. 1972. Regeneration and changes in synaptic connections between individual nerve cells in the central nervous system of the leech. *Proc. Natl. Acad. Sci.* **69**:636–639. (10)

―――――. 1973. Conductance changes, an electrogenic pump and the hyperpolarization of leech neurones following impulses. *J. Physiol.* **229**:635–665. (5,6,10)

Jansen, J.K.S., K.J. Muller, and J.G. Nicholls. 1974. Persistent modification of snyaptic interactions between sensory and motor nerve cells following discrete lesions in the central nervous system of the leech. *J. Physiol.* **242**:289–305. (6,10)

Johnson, J.R. 1816. *A treatise on the medicinal leech.* Longman, Hurst, Rees, Orme and Browne, London. (3)

―――――. 1825. *Further observations on the medicinal leech.* Longman, Hurst, Rees, Orme, Browne and Green. London. (3)

Kandel, E.R. 1976. *Cellular basis of behavior: An introduction to behavioral neurobiology.* W.H. Freeman, San Francisco. (6,10)

Kandel, E.R., T.J. Carew, and J. Koester. 1976. Principles relating to the biophysical properties of neurons and their patterns of interconnections to behavior. In *Electrobiology of nerve synapse and muscle* (ed. J.B. Reuben et al.), pp. 187-215. Raven Press, New York. (7)

Katz, B. and R. Miledi. 1970. Membrane noise produced by acetylcholine. *Nature* **226**:962–963. (6)

―――――. 1972. The statistical nature of the acetylcholine potential and its molecular components. *J. Physiol.* **224**:665–699. (6)

Kehoe, J. 1972. Three acetylcholine receptors in *Aplysia* neurones. *J. Physiol.* **225**:115–146. (8)

———. 1972. The physiological role of three acetylcholine receptors in synaptic transmission in *Aplysia*. *J. Physiol.* **225**:147–172. (6)

Kerkut, G.A., C.B. Sedden, and R.J. Walker. 1966. The effect of DOPA, methyl DOPA and reserpine on the dopamine content of the brain of the snail, *Helix aspersa*. *Comp. Biochem. Physiol.* **18**:921–930. (8)

———. 1967. Cellular localization of monoamines by fluorescence microscopy in *Hirudo medicinalis* and *Lumbricus terrestris*. *Comp. Biochem. Physiol.* **21**:687–690. (8)

———. 1967. A fluorescence microscopic and electrophysiological study of the giant neurones of the ventral nerve cord of *Hirudo medicinalis*. *J. Physiol.* **189**:83–85P. (8)

Keyser, K.T. 1979. "Leydig cells within the C.N.S. of the leech." Ph.D. thesis, State University of New York, Stony Brook. (D)

Keyser, K.T., B.M. Frazer, and C.M. Lent. 1981. Physiological and anatomical properties of Leydig cells in the segmental nervous system of the leech. *J. Comp. Physiol.* (in press). (D)

Kleinhaus, A.L. and J.W. Prichard. 1974. Electrophysiological properties of the leech subesophageal ganglion. *Brain Res.* **72**:332–336.

———. 1975. Calcium-dependent action potentials produced in leech Retzius cells by tetraethylammonium chloride. *J. Physiol.* **246**:351–361. (6)

———. 1977. Close relation between TEA responses and Ca-dependent membrane phenomena of four identified leech neurones. *J. Physiol.* **270**:181–194. (6)

Klemm, D.J. 1975. Studies on the feeding relationships of leeches (Annelida: Hirudinea) as natural associates of mollusks. *Sterkiana* **58**:1–50 and **59**:1–20. (2)

———. 1981. Freshwater leeches (Annelida: Hirudinea) of North America. In *Biota of freshwater ecosystems identification manual.* Superintendent of Documents, U.S. Government Printing Office, Washington, D.C. (2)

Kling, U. and G. Szekely. 1968. Simulation of rhythmic nervous activities. I. Function of networks with cyclic inhibitions. *Kybernetik* **5**:89–103. (7)

Kramer, A.P. 1981. The nervous system of the glossiphoniid leech *Haementeria ghilianii*. II. Synaptic pathways controlling body wall shortening. *J. Comp. Physiol.* (in press). (6)

Kramer, A.P. and J.R. Goldman. 1981. The nervous system of the glossiphoniid leech *Haementeria ghilianii*. I. Identification of neurons. *J. Comp. Physiol.* (in press). (2,5)

Kravitz, E.A., B.A. Battelle, P.D. Evans, B.R. Talamo, and B.G. Wallace. 1976. Octopamine neurons in lobsters. *Neurosci. Symp.* **1**:67–81. (8)

Kretz, J.R., G.S. Stent, and W.B. Kristan, Jr. 1976. Photosensory input pathways in the medicinal leech. *J. Comp. Physiol.* **106**:1–37. (4,5,D)

Kristan, W.B., Jr. 1980. The generation of rhythmic motor patterns. In *Information processing in the nervous system* (ed. H. Pinsker and W.D. Willis), pp. 241–261. Raven Press, New York. (7)

Kristan, W.B. and R.L. Calabrese. 1976. Rhythmic swimming activity in neurons of the isolated nerve cord of the leech. *J. Exp. Biol.* **65**:643–668. (6,7)

Kristan, W.B., Jr., and P.B. Guthrie. 1977. Acquisition of swimming behavior in chronically isolated single segments of the leech. *Brain Res.* **131**:191–195. (10)

Kristan, W.B., Jr. and G.S. Stent. 1976. Peripheral feedback in the leech swimming rhythm. *Cold Spring Harbor Symp. Quant. Biol.* **40**:663–674. (5)

Kristan, W.B., Jr., S.J. McGirr, and G.V. Simpson. 1981. Behavioral and mechanosensory neuron responses to skin stimulation in leeches. *J. Exp. Biol.* (in press). (7)

Kristan, W.B., G.S. Stent, and C.A. Ort. 1974. Neuronal control of swimming in the medicinal leech. I. Dynamics of the swimming rhythm. *J. Comp. Physiol.* **94**:97–119. (4,5,7)

―――――. 1974. Neuronal control of swimming in the medicinal leech. III. Impulse patterns of motor neurons. *J. Comp. Physiol.* **94**:155–176. (7)

Kuffler, D.P. 1978. Neuromuscular transmission in longitudinal muscle of the leech, *Hirudo medicinalis*. *J. Comp. Physiol.* **124**:333–338. (6,8)

Kuffler, D.P. and K.J. Muller. 1974. The properties and connections of supernumerary sensory and motor nerve cells in the central nervous system of an abnormal leech. *J. Neurobiol.* **5**:331–348. (5)

Kuffler, S.W. and J.G. Nicholls. 1966. The physiology of neuroglial cells. *Ergeb. Physiol. Biol. Chem. Exp. Pharmakol.* **57**:1–90. (4,8)

Kuffler, S.W. and D.D. Potter. 1964. Glia in the leech nervous system: Physiological properties and neuron-glia relationships. *J. Neurophysiol.* **27**:290–320. (4,6)

Kupfermann, I. 1979. Modulatory actions of neurotransmitters. *Annu. Rev. Neurosci.* **2**:447–465. (8)

Kuzmina, L.V. 1968. Distribution of biogenic monoamines in the nervous system of the body segments of the leech, *Hirudo medicinalis*. *J. Evol. Biochem. Physiol.* (Suppl., Physiology and biochemistry of invertebrates), pp. 50–56. (8)

Lankester, E.R. 1880. On the connective and vasifactive tissue of the medicinal leech. *Q. J. Microsc. Sci.* **20**:307–317. (4)

Larsen, W.J. 1977. Structural diversity of gap junctions. *Tissue Cell* **9**:373–394. (6)

Le Douarin, N.M. and M.-A.M. Teillet. 1974. Experimental analysis of the migration and differentiation of neuroblasts of the autonomic nervous system and of neuroectodermal mesenchymal derivatives, using biological cell marking technique. *Dev. Biol.* **41**:162–184. (8)

Lent, C.M. 1973. Retzius cells: Neuro effectors controlling mucus release by the leech. *Science* **179**:693–696. (8,D)

―――――. 1981. Morphology of neurons containing monoamines within leech segmental ganglia. *J. Exp. Zool.* **216**:311–316. (D)

Lent, C.M. and B.M. Frazer. 1977. Connectivity of the monoamine-containing neurones in central nervous system of leech. *Nature* **266**:844–847. (D)

Lent, C.M., J. Ono, K.T. Keyser, and H. Karten. 1979. Identification of serotonin within vital-stained neurons from leech ganglia. *J. Neurochem.* **32**:1559–1563. (8, D)

Levitan, H. and L. Tauc. 1972. Acetylcholine receptors: Topographic distribution and pharmacological properties of two receptor types of a single molluscan neuron. *J. Physiol.* **222**:537–558. (6)

Livanov, N. 1904. Untersuchungen zur Morphologie der Hirudineen. II. Das Nervensystem des Vorderen Körperendes und seine Metamerie. *Zool. Jahrb. Anat.* **20**:153–226. (4)

Livingstone, M.S., R.M. Harris-Warrick, and E.A. Kravitz. 1980. Serotonin and octopamine produce opposite postures in lobsters. *Science* **208**:76–79. (8)

Louis, P.C.A. 1836. *Researches on the effect of blood letting in some inflammatory diseases and on the influence of tartarised antimony and vesication in pneumonias* (Translated by C.G. Putnam). Hilliard Gray, Boston. (3)

Lowe, P. 1634. *A discourse on the whole art of chirurgerie.* Book 9, chapter 13, pp. 391–393. T. Purfoot, London. (3)

Macagno, E.R. 1980. Number and distribution of neurons in the leech segmental ganglion. *J. Comp. Neurol.* **190**:283–302. (4,6)

Macagno, E.R., K.J. Muller, W. Kristan, S. DeRiemer, R. Stewart, and B. Granzow. 1981. Mapping of neuronal contacts with intracellular injection of horseradish peroxidase and Lucifer Yellow in combination. *Brain Res.* **217**:143–149. (6)

MacDonald, J.F., J.L. Barker, D.L. Gruol, L.M. Huang, and T.G. Smith. 1979. Desensitizing excitatory responses to peptides, purines, protons and drugs revealed using cultured mammalian neurons. *Soc. Neurosci. Symp.* **5**:592. (Abstr.). (8)

MacIntosh, F.C. and W.L.M. Perry. 1950. Biological estimation of ACh. In *Methods in medical research* (ed. R. W. Gerard), pp. 78-92. Year Book, Chicago. (8)

Magni, F. and M. Pellegrino. 1978. Neural mechanisms underlying the segmental and generalized cord shortening reflexes in the leech. *J. Comp. Physiol.* **124**:339–351. (D)

————. 1978. Patterns of activity and the effects of activation of the fast conducting system on the behavior of unrestrained leeches. *J. Exp. Biol.* **76**:123–135. (6,D)

Mann, K.H. 1953. The segmentation of leeches. *Biol. Rev.* **28**:1–15. (4)

————. 1961. *Leeches (Hirudinea).* Pergamon Press, New York. (8)

————. 1962. *Leeches (Hirudinea). Their structure, physiology, ecology and embryology.* Pergamon Press, New York. (3,4,7)

Manton, S.M. 1972. The evolution of arthropodal locomotory mechanisms, Part 10. *J. Linn. Soc. Lond. Zool.* **51**:203–400. (2)

Markwardt, F. 1957. Die Isolierung und Chemische Charakterisirung des Hirudin. *Hoppe-Seyler's Z. Physiol. Chem.* **308**:147–156. (3)

Marsden, C.A. and G.A. Kerkut. 1969. Fluorescent microscopy of the 5-HT and catecholamine-containing cells in the central nervous system of the leech *Hirudo medicinalis*. *Comp. Biochem. Physiol.* **31**:851–862. (8)

Mason, A. and L.D. Leake. 1978. Morphology of leech Retzius cells demonstrated by intracellular injection of horseradish peroxidase. *Comp. Biochem. Physiol.* **61A**:213–216. (D)

Mason, A., A.J. Sunderland, and L.D. Leake. 1979. Effects of leech Retzius cells on body wall muscles. *Comp. Biochem. Physiol.* **63C**:359–361. (6,8,D)

Massoulié, J,, S. Bon, and M. Vigny. 1980. The polymorphism of cholinesterase in vertebrates. *Neurochem. Int.* **2**:161–184. (8)

McAdoo, D.J. and R.E. Coggeshall. 1976. Gas chromatographic-mass spectrometric analysis of biogenic amines in identified neurons and tissues of *Hirudo medicinalis*. *J. Neurochem.* **26**:163–167. (8)

McCaman, M.W., D. Weinreich, and R.E. McCaman. 1973. The determination of picomole levels of 5-hydroxytryptamine and dopamine in *Aplysia*, *Tritonia* and leech nervous tissue. *Brain Res.* **53**:129–137. (8)

McDougall, W. 1903. The nature of inhibitory processes within the nervous system. *Brian* **26**:153–191. (7)

McMahan, U.J. and S.W. Kuffler. 1971. Visual identification of synaptic boutons on living ganglion cells and of varicosities in postganglionic axons in the heart of the frog. *Proc. R. Soc. Lond. B* **177**:485–508. (6)

Meech, R.W. and N.B. Standen. 1975. Potassium activation in *Helix aspersa* under voltage clamp: A component mediated by calcium influx. *J. Physiol.* **249**:211-239. (7)

Merickel, M. and R. Gray. 1980. Investigation of burst generation by the electrically coupled cyberchron network in the snail *Helisoma* using a single electrode voltage clamp. *J. Neurobiol.* **11**:73–102. (7)

Miller, J.P. and A. Selverston. 1979. Rapid killing of single neurons by irradiation of intracellularly injected dye. *Science* **206**:702–704. (A)

Minz, B. 1932. Pharmakologische Untersuchungen am Blutegelpraeparat: Zugleich eine Methode zum biologischen Nachweis von Acetylcholine bei Arwesenheit anderer pharmakologisch wirksamer koerpereigner Stoffe. *Naunyn-Schmeiderbergs Arch. Pathol. Pharmakol.* **168**:292–304. (3)

Miyazaki, S. and J.G. Nicholls. 1976. The properties and connexions of nerve cells in leech ganglia maintained in culture. *Proc. R. Soc. Lond. B* **194**:295–311. (6,10)

Miyazaki, S., J.G. Nicholls, and B.G. Wallace. 1976. Modification and regeneration of synaptic connections in cultured leech ganglia. *Cold Spring Harbor Symp. Quant. Biol.* **40**:483–493. (10)

Moquin-Tandon, A. 1846. *Monographie sur la famille des hirudinees, atlas*. J.B. Bailliere, Paris. (3)

Muller, K.J. 1979. Synapses between neurones in the central nervous system of the leech. *Biol. Rev.* **54**:99–134. (5,6,10)

Muller, K.J. and S. Carbonetto. 1977. Two ways that an electrical connection is re-established in the leech. *Neurosci. Abstr.* **3**:353. (10)

――――. 1979. The morphological and physiological properties of a regenerating synapse in the C.N.S. of the leech. *J. Comp. Neurol.* **185**:485–516. (6,10,D)

Muller, K.J. and U.J. McMahan. 1975. The arrangement and structure of synapses formed by specific sensory and motor neurons in segmental ganglia of the leech. *Anat. Rec.* **181**:432. (6)

――――. 1976. The shapes of sensory and motor neurones and the distribution of their synapses in ganglia of the leech: A study using intracellular injection of horseradish peroxidase. *Proc. R. Soc. Lond. B* **194**:481–499. (5,6,8,D)

Muller, K.J. and J.G. Nicholls. 1974. Different properties of synapses between a single sensory neurone and two different motor cells in the leech C.N.S. *J. Physiol.* **238**:357–369. (6)

Muller, K.J. and S.A. Scott. 1979. Correct axonal regeneration after target cell removal in the central nervous system of the leech. *Science* **206**:87–89. (10)

――――――. 1980. Removal of the synaptic target permits terminal sprouting of a mature intact axon. *Nature* **283**:89–90. (10)

――――――. 1981. Transmission at a "direct" electrical connexion mediated by an interneurone in the leech. *J. Physiol.* **311**:565–584. (5,6,A,D)

Muller, K.J., S.A. Scott, and B.E. Thomas. 1978. Specific associations between sensory cells. *Carnegie Inst. Wash. Yearbook* **77**:69–70. (8)

Nicholls, J.G. and D.A. Baylor. 1968. Specific modalities and receptive fields of sensory neurons in the CNS of the leech. *J. Neurophysiol.* **31**:740–756. (4,5,D)

――――――. 1968. Long lasting hyperpolarization after activity of neurons in the leech central nervous system. *Science* **162**:279–281. (4)

Nicholls, J.G. and D. Purves. 1970. Monosynaptic chemical and electrical connexions between sensory and motor cells in the central nervous system of the leech. *J. Physiol.* **209**:647–667. (6,A)

――――――. 1972. A comparison of chemical and electrical synaptic transmission between single sensory cells and a motorneurone in the central nervous system of the leech. *J. Physiol.* **225**:637–656. (6)

Nicholls, J.G. and D. Van Essen. 1974. The nervous system of the leech. *Sci. Am.* **230**:38–48. (4)

Nicholls, J.G. and B.G. Wallace. 1978. Quantal analysis of transmitter release at an inhibitory synapse in the C.N.S. of the leech. *J. Physiol.* **281**:171–185. (6)

――――――. 1978. Modulation of transmission at an inhibitory synapse in the central nervous system of the leech. *J. Physiol.* **281**:157–170. (6,7)

Nicholls, J.G. and D.E. Wolfe. 1967. Distribution of ^{14}C-labeled sucrose, insulin, and dextran in extracellular spaces and in cells of the leech central nervous system. *J. Neurophysiol.* **30**:1574–1592. (4)

Orchard, I. and R. Webb. 1980. The projections of neurosecretory cells in the brain of the North-American medicinal leech, *Macrobdella decora*, using intracellular injection of horseradish peroxidase. *J. Neurobiol.* **11**: 229–242. (4,D)

Ort, C.A., W.B. Kristan, and G.S. Stent. 1974. Neuronal control of swimming in the medicinal leech. II. Identification and connections of motor neurons. *J. Comp. Physiol.* **94**:121–154. (4,5,6,7,C,D)

Osborne, N.N., G. Briel, and V. Neuhoff. 1972. The amine and amino acid composition in the Retzius cells of the leech *Hirudo medicinalis*. *Experentia* **28**:1015–1018. (8)

Osler, W. 1892. *The principle and practice of medicine*. Young and Pertland, London. (3)

Pappas, G.D., Y. Asada, and M.V.L. Bennett. 1971. Morphological correlates of increased coupling resistance at an electronic synapse. *J. Cell Biol.* **49**:173–188. (6)

Paré, A. 1678. *The works of Ambroise Paré*, 2nd edition. (Translated by Johnson.) M. Clark (printer), London. (3)

Parnas, I. and D. Bowling. 1977. Killing of single neurons by intracellular injection of proteolytic enzymes. *Nature* **370**:626–628. (6)

Parnas, I., S. Hochstein, and H. Parnas. 1976. Theoretical analysis of parameters leading to frequency modulation along an inhomogeneous axon. *J. Neurophysiol.* **39**:909–923. (10)

Parnas, I., M.E. Spira, R. Werman, and F. Bergman. 1969. Nonhomogeneous conduction in giant axons of the nerve cord of *Periplaneta americana*. *J. Exp. Biol.* **50**:635–649. (10)

Payne, J.F. 1904. *English medicine in Anglo-Saxon times*. Clarendon, Oxford. (3)

Payton, B.W. and W.R. Loewenstein. 1968. Stability of electrical coupling in leech giant nerve cells: Divalent cations, propionate ions, tonicity and pH. *Biochim. Biophys. Acta* **150**:156–158. (6)

Peracchia, C. 1973. Low resistance junctions in crayfish. II. Structural details and further evidence for intercellular channels by freeze-fracture and negative-staining. *J. Cell Biol.* **57**:66–76. (6)

———. 1973. Low resistance junction in crayfish. I. Two arrays of globules in junctional membranes. *J. Cell Biol.* **57**:54–65. (6)

Perez, H.V.Z. 1942. On the chromaffin cells of the nerve ganglia of *Hirudo medicinalis* Lin. *J. Comp. Neurol.* **76**:367–401. (8)

Perkel, D.H. and B. Mulloney. 1974. Motor pattern production in reciprocally inhibitory neurons exhibiting postinhibitory rebound. *Science* **185**:181–183. (7)

Peterson, E.L. and R.L. Calabrese. 1981. Dynamic analysis of a rhythmic neural circuit in the leech *Hirudo medicinalis*. *J. Neurophysiol.* (in press). (7)

Phillips, C.E. and W.O. Friesen. 1981. The fine structure of leech sensilla. *Neurosci. Abstr.* **7**:00. (in press). (5)

Pitman, R.M., C.D. Tweedle, and M.J. Cohen. 1972. Electrical responses of insect central neurons: Augmentation by nerve section or colchicine. *Science* **178**:507–509. (10)

Poll, H. and A. Sommer. 1903. Ueber phaeochrome Zellen in Centralnervensystem des Blutegels. *Arch. Anat. Physiol. (Physiol. Abt.)* 1903 (Nr. X). (8)

Poon, M. 1976. "A neuronal network generating the swimming rhythm in the leech." Ph.D. thesis, University of California, Berkeley. (7)

Poon, M., W.O. Friesen, and G.S. Stent. 1978. Neuronal control of swimming in the medicinal leech. V. Connections between the oscillatory interneurons and the motor neurons. *J. Exp. Biol.* **75**:43–63. (7)

Potter, L.T. 1970. Synthesis, storage and release of (^{14}C) acetylcholine in isolated rat diaphragm muscles. *J. Physiol.* **206**:145–166. (8)

Purves, D. and U.J. McMahan. 1972. The distribution of synapses on a physiologically identified motor neuron in the central nervous system of the leech. *J. Cell Biol.* **55**:205–220. (6)

———. 1973. Procion Yellow as a marker for electron microscopic examination of functionally identified nerve cells. In *Intracellular staining in neurobiology* (ed. S.B. Kater and C. Nicholson), pp. 72–81. Springer-Verlag, New York. (6)

Rakic, P. 1975. Local circuit neurones. *Neurosci. Res. Prog. Bull.* **13**:291–440. (6)

Ramon y Cajal, S. 1928. *Degeneration and regeneration of the nervous system.* (Translated and edited by R.M. May.) Hofner, New York. (10)

Ready, D. and J. Nicholls. 1979. Identified neurones isolated from leech CNS make selective connections in culture. *Nature* **281**:67–69. (8,10)

Retzius, G. 1891. Zur Kenntniss des centralen Nervensystems der Wurmer. *Biologische Untersuchungen, Neue Folge II*, 1–28. Samson and Wallin, Stockholm. (4,5)

Rieske, E., P. Schubert, and G.W. Kreutzberg. 1975. Transfer of radioactive material between electrically coupled neurones of the leech central nervous system. *Brain Res.* **84**:365–382. (6)

Rolleston, J.D. 1939. F.J.V. Broussais (1771-1838): His life and doctrines. *Proc. R. Soc. Med.* **32**:405–413. (3)

Rosenbluth, J. 1973. Postjunctional membrane specialization at cholinergic myoneural junctions in the leech. *J. Comp. Neurol.* **151**:399–405. (6)

Rubin, E. 1978. The caudal ganglion of the leech, with particular reference to homologues of segmental touch receptors. *J. Neurophysiol.* **9**:393–405. (4)

Rude, S. 1969. Monoamine-containing neurons in the central nervous system and peripheral nerves of the leech, *Hirudo medicinalis. J. Comp. Neurol.* **136**:349–371. (5,8)

Rude, S., R.E. Coggeshall, and L.S. Van Orden, III. 1969. Chemical and ultrastructural identification of 5-hydroxytryptamine in an identified neuron. *J. Cell Biol.* **41**:832–854. (8)

Salzberg, B.M., H.V. Davila, and L.B. Cohen. 1973. Optical recording of impulses in individual neurones of an invertebrate central nervous system. *Nature* **246**:508–509. (5)

Sargent, P.B. 1975. "Transmitters in the leech central nervous system: Analysis of sensory and motor cells." Ph.D. thesis, Harvard University, Cambridge, Massachusetts. (8,D)

————. 1977. Synthesis of acetylcholine by excitatory motoneurons in central nervous system of the leech. *J. Neurophysiol.* **40**:453–460. (8)

————. 1977. Transmitters in the leech central nervous system: Analysis of sensory and motor cells. *J. Neurophysiol.* **40**:453–460. (5)

Sargent, P.B., K.-W. Yau, and J.G. Nicholls. 1977. Extrasynaptic receptors on cell bodies of neurons in the central nervous system of the leech. *J. Neurophysiol.* **40**:446–452. (5,6,8,10)

Sawada, M. and R.E. Coggeshall. 1976. Ionic mechanism of 5-hydroxytryptamine induced hyperpolarization and inhibitory junctional potential in body wall muscle cells of *Hirudo medicinalis. J. Neurobiol.* **7**:63–73. (8)

Sawada, M., J.M. Wilkinson, D.J. McAdoo, and R.E. Coggeshall. 1976. The identification of two inhibitory cells in each segmental ganglion of the leech and studies on the ionic mechanism of the inhibitory junctional potentials produced by these cells. *J. Neurobiol.* **7**:435–445. (8,D)

Sawyer, R.T. 1970. Observations on the natural history and behavior of *Erpobdella punctata* (Leidy) (Annelida: Hirudinea). *Am. Mid. Nat.* **83**:65–80. (2)

──────── . 1971. The phylogenetic development of brooding behavior in the Hirudinea. *Hydrobiologia* **37**:197–204. (2)

──────── . 1972. *North American freshwater leeches, exclusive of the Piscicolidae, with a key to all species.* University of Illinois Press, Urbana. (2)

──────── . 1974. Ecology of freshwater leeches. In *Pollution ecology of freshwater organisms* (ed. C.W. Hart and S.L.H. Fuller), pp. 81–142. Academic Press, New York. (2)

──────── . 1978. Domestication of the world's largest leech for developmental neurobiology. *Year Book of the American Philosophical Society*, pp. 212–213. (2)

──────── . 1981. Terrestrial Leeches. In *Soil biology guide* (ed. D.L. Dindal). Interscience, New York. (2)

──────── . 1981. An expedition to Borneo to study the aggressive behavior of land leeches, with collateral analysis of their anti-coagulants for medicinal purposes. *Year Book of the American Philosophical Society.* (In press.)

Sawyer, R.T. and S. Fitzgerald. 1981. Leech circulatory system. In *Invertebrate blood cells* (ed. N.A. Radcliffe and A.F. Rowley), vol. 1, pp. 141–159. Academic Press, London. (2)

Sawyer, R.T. and D.L. Hammond. 1973. Distribution, ecology and behavior of the marine leech *Calliobdella carolinensis* (Annelida: Hirudinea), parasitic on the Atlantic menhaden in epizootic proportions. *Biol. Bull.* **143**:373–388. (2)

Sawyer, R.T. and R.H. Shelley. 1976. New records and species of leeches (Annelida: Hirudinea) from North and South Carolina. *J. Nat. Hist.* **10**:65–97. (2)

Sawyer, R.T., A.R. Lawler, and R.H. Overstreet. 1975. The marine leeches of the eastern United States and the Gulf of Mexico with a key to the species. *J. Nat. Hist.* **9**:633–667. (2)

Sawyer, R.T., F. LePont, D.K. Stuart, and A.P. Kramer. 1981. Growth and reproduction of the giant glossiphoniid leech *Haementeria ghilianii*. *Biol. Bull.* **160**:322–331. (2,9)

Schleip, W. 1936. Ontogenie der Hirudineen. *In Klassen und Ordnungen des Tierreichs* (ed. H.G. Bronn), vol. 4, Div. III, Book 4, Part 2, pp. 1-121. Akademie Verlagsgesellschaft, Leipzig. (9)

Schmucker, J.L. 1776. Historisch-practische Abhandlung vom medicinischen Gebrauche der Blutegel. *Verm. Chir. Schrift. Berl.* **1**:75–116. (3)

Schnaitman, C., V.G. Erwin, and J.W. Greenawalt. 1967. The submitochondrial localization of monoamine oxidase. An enzymatic marker for the outer membrane of rat liver mitochondria. *J. Cell Biol.* **32**:719–735. (8)

Schneider, F.H. and C.N. Gillis. 1965. Catecholamine biosynthesis *in vivo*: An application of thin layer chromatography. *Biochem. Pharmacol.* **14**:623–626. (8)

Scott, S.A. and K.J. Muller. 1980. Synapse regeneration and signals for directed axonal growth in the C.N.S. of the leech. *Dev. Biol.* **80**:345–363. (6,10,A)

Shafer, M.R. and R.L. Calabrese. 1981. Similarities and differences in the structure of segmentally homologous neurons that control the hearts in the leech, *Hirudo medicinalis*. *Cell Tissue Res.* **214**:137–153. (D)

Smith, T.G., Jr., J.L. Barker, and H. Gainer. 1975. Requirements for bursting pacemaker activity in molluscan neurons. *Nature* **253**:450–452. (7)

Staehelin, L.A. 1974. Structure and function of intercellular junctions. *Int. Rev. Cytol.* **39**:191–283. (6)

Stent, G.S., W.J. Thompson, and R.L. Calabrese. 1979. Neural control of heartbeat in the leech and in some other invertebrates. *Physiol. Rev.* **59**: 101–136. (7,D)

Stent, G.S., D.A. Weisblat, S.S. Blair, and S.L. Zackson. 1982. Cell lineage in the development of the leech nervous system. In *Neuronal development* (ed. N. Spitzer). Plenum Press, New York. (In press.) (9)

Stent, G.S., W.B. Kristan, Jr., W.O. Friesen, C.A. Ort, M. Poon, and R.L. Calabrese. 1978. Neuronal generation of the leech swimming movement. *Science* **200**:1348–1357. (2,6,7,D)

Stewart, W.W. 1978. Intracellular marking of neurons with a highly fluorescent naphthalimide dye. *Cell* **14**:741–759. (6,D)

―――――. 1981. Lucifer dyes — Highly fluorescent dyes for biological tracing. *Nature* **292**:17–21. (6,C)

Strumwasser, F. 1967. Types of information stored in single neurons. In *Invertebrate nervous systems* (ed. C.A.G. Wiersma), pp. 219–319. University of Chicago Press, Chicago. (7)

―――――. 1971. The cellular basis of behavior in *Aplysia*. *J. Psychiatr. Res.* **8**:237–289. (7)

Stuart, A.E. 1970. Physiological and morphological properties of motoneurones in the central nervous system of the leech. *J. Physiol.* **209**:627–646. (4,5,6,7,A,C,D)

Stuart, A.E., A.J. Hudspeth, and Z.W. Hall. 1974. Vital staining of specific monoamine-containing cells in the leech central nervous system. *Cell Tissue Res.* **153**:55–61. (6,8,D)

Suicide. 1892. Note in *N.Y. Med. J.* **56**:103. (3)

Sunderland, A.J. 1980. A hitherto undocumented pair of neurons in the segmental ganglion of the leech which receive synaptic input from mechanosensory cells. *Comp. Biochem. Physiol.* **67A**:299–302. (D)

Szekely, G. 1967. Development of limb movements: Embryological, physiological and model studies. *Ciba Found. Symp. on growth of the nervous system* (eds. G. Wolstenholme and M. O'Connor), pp. 77–93. Little Brown and Co., Boston. (7)

Szerb, J.C. 1961. The estimation of acetylcholine using leech muscle in a microbath. *J. Physiol.* **158**:8P–9P. (3)

Thomas, R. 1825. *The modern practice of physic*, 8th edition. Collins and Collins, New York. (3)

Thompson, W. and G.S. Stent. 1976. Neuronal control of heartbeat in the medicinal leech. I. Generation of the vascular constriction rhythm by heart motor neurons. *J. Comp. Physiol.* **111**:261–279. (4,7,C,D)

———. 1976. Neuronal control of heartbeat in the medicinal leech. II. Intersegmental coordination of heart motor neuron activity by heart interneurons. *J. Comp. Physiol.* **111**:281–307. (6,7,10,D)

———. 1976. Neuronal control of heartbeat in the medicinal leech. III. Synaptic relations of the heart interneurons. *J. Comp. Physiol.* **111**:309–333. (6,7)

Tisdale, A.D. and Y. Nakajima. 1976. Fine structure of synaptic vesicles in the two types of nerve terminals in crayfish stretch receptor organs: Influence of fixation methods. *J. Comp. Neurol.* **165**:369–386. (6)

Tulsi, R.S. and R.E. Coggeshall. 1971. Neuromuscular junctions on the muscle cells in the central nervous system of the leech, *Hirudo medicinalis*. *J. Comp. Neurol.* **141**:1–16. (6)

Van Essen, D.C. 1973. The contribution of membrane hyperpolarization to adaptation and conduction block in sensory neurones of the leech. *J. Physiol.* **230**:509–534. (5,6,10)

Van Essen, D.C. and J.K.S. Jansen. 1977. The specificity of re-innervation by identified sensory and motor neurons in the leech. *J. Comp. Neurol.* **171**:433–454. (5,10,A,D)

Viallia, M. 1934. Le cellule cromaffini dei gangli nervosa negli Irudinei. *Atti. Soc. Ital. Sci. Nat.* **73**:57–73. (8)

Vitet, L. 1809. *Traite de la sangsue medicinale*. Vitet, Paris. (3)

Von der Wense, T. 1939. Ueber den nachweis von Adrenalin in Wurmern and Insekten. *Pfluegers Arch.* **241**:284–288. (8)

von Uexküll, J. 1905. Studien veber den Tonus III. Die Blutegel. *Z. Biol.* **46**:(N.F. 28):372–402. (7)

Walker, R.J., G.N. Woodruff, and G.A. Kerkut. 1968. The effect of acetylcholine and 5-hydroxytryptamine on electrophysiological recordings from muscle fibres of the leech *Hirudo medicinalis*. *Comp. Biochem. Physiol.* **24**:987–990. (8)

———. 1970. The action of cholinergic antagonists on spontaneous excitatory potentials recorded from the body wall of the leech, *Hirudo medicinalis*. *Comp. Biochem. Physiol.* **32**:690–701. (8)

Wallace, B.G. 1976. The biosynthesis of octopamine—Characterization of lobster tyramine β-hydroxylase. *J. Neurochem.* **26**:761–770. (8)

———. 1980. Selective neurite atrophy during development of cells in the leech C.N.S. *Neurosci. Abstr.* **6**:679. (5)

———. 1981. Distribution of AChE in cholinergic and non-cholinergic neurons. *Brain Res.* (in press). (8)

Wallace, B.G., M. Adal, and J.G. Nicholls. 1977. Regeneration of synaptic connexions of sensory neurones in leech ganglia in culture. *Proc. R. Soc. Lond. B* **199**:567–585. (10,A)

Wallace, B.G., B.R. Talamo, P.D. Evans, and E.A. Kravitz. 1974. Octopamine: Selective association with specific neurons in the lobster nervous system. *Brain Res.* **74**:349–355. (8)

Watanabe, A., S. Obara, and T. Akiyama. 1967. Pacemaker potentials for the periodic burst discharge in the heart ganglion of a stomatopod, *Squilla oratoria*. *J. Gen. Physiol.* **50**:839–862. (7)

Weeks, J.C. 1980. "The roles of identified interneurons in initiating and generating the swimming motor pattern of leeches." Ph.D. thesis, University of California, San Diego. (7,D)

———. 1981. Neuronal basis of leech swimming: Separation of swim initiation, pattern generation and intersegmental coordination by selective lesions. *J. Neurophysiol.* **45**:698–723. (7)

Weeks, J.C. and W.B. Kristan, Jr. 1978. Initiation, maintenance and modulation of swimming in the medicinal leech by the activity of a single neuron. *J. Exp. Biol.* **77**:71–88. (7,D)

Weisblat, D.A., R.T. Sawyer, and G.S. Stent. 1978. Cell lineage analysis by intracellular injection of a tracer enzyme. *Science* **202**:1295–1298. (9)

Weisblat, D.A., G. Harper, G.S. Stent, and R.T. Sawyer. 1980. Embryonic cell lineages in the nervous system of the glossiphoniid leech *Helobdella triserialis*. *Dev. Biol.* **76**:58–78. (9)

Weisblat, D.A., S.L. Zackson, S.S. Blair, and J.D. Young. 1980. Cell lineage analysis by intracellular injection of fluorescent tracers. *Science* **209**:1538–1541. (9,A)

Weiss, K.R., J.L. Cohen, and I. Kupfermann. 1978. Modulatory control of buccal musculature by a serotonergic neuron (metacerebral cell) in *Aplysia*. *J. Neurophysiol.* **41**:181–203. (8)

Whitman, C.O. 1892. The metamerism of *Clepsine*. In *Festschrift zum 70, Geburtstage R. Leuckarts*, pp. 385–395. Engelmann, Leipzig. (9)

———. 1878. The embryology of *Clepsine*. *Q. J. Micros. Sci.* **18**:215–315. (9)

———. 1887. A contribution to the history of germ layers in *Clepsine*. *J. Morphol.* **1**:105–182. (9)

Willard, A.L. 1981. Effects of serotonin on the generation of the motor program for swimming by the medicinal leech. *J. Neurosci.* (in press). (7,8,D)

Wolfe, D.E. and J.G. Nicholls. 1967. Uptake of radioactive glucose and its conversion to glycogen by neurons and glial cells in the leech central nervous system. *J. Neurophysiol.* **30**:1593–1609. (4)

Wood, M.R., K.H. Pfenninger, and M.J. Cohen. 1977. Two types of presynaptic configurations in insect central synapses: An ultrastructural analysis. *Brain Res.* **130**:25–45. (6)

Wrona, F.J., R.W. Davies, L. Linton, and J. Wilkialis. 1981. Competition and coexistence between *Glossiphonia complanata* and *Helobdella stagnalis* (Glossiphoniidae: Hirudinoidea). *Oecologia* **48**:133–137. (2)

Yaksta-Sauerland, B.A. and R.E. Coggeshall. 1973. Neuromuscular junctions in the leech. *J. Comp. Neurol.* **151**:85–99. (6,8)

Yanagisawa, H. and E. Yokoi. 1938. The purification of hirudin and action principle of *Hirudo medicinalis*. *Proc. Imp. Acad. Tokyo.* **14**:69–70. (4)

Yau, K.W. 1975. "Receptive fields, geometry and conduction block of sensory cells in the leech central nervous system." Ph.D. thesis, Harvard University, Cambridge, Massachusetts. (6,D)

———. 1976. Receptive fields, geometry and conduction block of sensory neurones in the CNS of the leech. *J. Physiol.* **263**:513–538. (5,6,10,C,D)

———. 1976. Physiological properties and receptive fields of mechanosensory neurones in the head ganglion of the leech: Comparison with homologous cells in segmental ganglia. *J. Physiol.* **263**:489–512. (4,5,D)

Young, J.O. and J.W. Ironmonger. 1979. The natural diet of *Erpobdella octoculata* (L.) (Hirudinea: Erpobdellidae) in British lakes. *Archiv fur Hydrobiologie* **87**:483–503. (2)

Young, S.R., R.D. Dedwyler, II, and W.O. Friesen. 1981. Responses of the medicinal leech to water waves. *J. Comp. Physiol.* (in press). (7)

Zipser, B. 1979. Voltage-modulated membrane resistance in coupled leech neurons. *J. Neurophysiol.* **42**:465–475. (5,6,D)

―――――. 1979. Identifiable neurons controlling penile eversion in the leech. *J. Neurophysiol.* **42**:455–464. (4,5,6,D)

―――――. 1980. Horseradish peroxidase nerve backfilling in leech. *Brain Res.* **182**:441–445. (D)

―――――. 1980. Identification of specific leech neurones immunoreactive to enkephalin. *Nature* **283**:857–858. (8,D)

Zipser, B. and R. McKay. 1981. Monoclonal antibodies distinguish identifiable neurones in the leech. *Nature* **289**:549–554. (5,D)

Name Index

Adal, M., 226, 234
Adam, A., 118, 143
Akert, K., 108
Akiyama, T., 146
Albucasis, 28, 33
Alving, B.O., 114, 143
Anderson, D.T., 7, 24
Andrew, W., 11, 24
Armett-Kibel, C., 82, 107
Asada, Y., 110
Ascher, P., 87, 107
Atwood, H.L., 102, 107
Autrum, H., 32, 33
Axelrod, J., 167

Bacq, Z.M., 156, 167
Bagdy, D., 9, 25
Barabas, E., 25
Barker, D.L., 169
Barker, J.L., 145, 170
Batelle, B.-A., 154, 167, 169

Baylor, D.A., 45, 48, 49, 52, 53, 55, 56, 59, 60, 65, 71, 72, 77, 78, 95, 97, 105, 107, 198, 199, 225, 281, 282, 287
Beeton, I., 32, 33
Bennett, M.V.L., 90, 94, 107, 110
Bergman, F., 226
Berry, M.S., 155, 167
Bianchi, S., 150, 151, 167
Biedl, A., 150, 167
Birks, R.I., 160, 167
Bittner, G.D., 102, 107, 234
Blackshaw, S.E., **51**, 63, 65, 70, 77, 154, 167
Blair, S.S., 193, 194, 195, 234
Blankenship, J.E., 162, 167
Boaistuau, P., 28, 33
Bon, S., 161, 167, 170
Boroffka, I., 118, 143
Bowling, D., 94, 107, 110, 228, 229, 231, 233

Italics indicate where full reference can be found; **boldface** type designates where author's article is located in this volume.

Name Index

Bownds, M.D., *169*
Boyle, M.B., *77*
Bradbury, S., 41, *49*
Bradley, G.W., 115, *144*
Briel, G., *170*
Brimijoin, S., 161, *167*
Brizuela, B.S., *25*
Brockbank, W., 29, *33*
Brown, T.G., 117, *144*
Brownstein, M.J., 160, *167*
Buchan, W., 31, *33*
Budzinski, A.Z., 9, *25*
Bullock, T.H., 113, *144*
Büsing, K.H., 39, *49*

Calabrese, R.L., *26*, 91, *109*, *111*, 121, 123, 125, 127, 134, 135, *144*, *145*, 217, *225*, 283, *287*
Camelli, E., 155, *167*
Carbonetto, S., 90, 95, *107*, *109*, 203, 207, 209, 211, 217, *225*, 226, 283, *286*
Carew, T.J., *144*
Carpenter, D.O., *167*
Cline, H., *182*
Clowes, W., 29, *33*
Cockayne, O. 28, *33*
Coggeshall, R.E., 43, *49*, 81, 87, 90, *107*, *111*, *144*, 149, 152, 153, 154, 155, 156, 159, 160, 162, *168*, *170*, *171*, *172*, *287*
Cohen, J.L., *111*, *172*
Cohen, L.B., *77*, *78*
Cohen, M.J., *226*
Coppée, G., 156, *167*
Cottrell, G.A., 160, *168*, *169*
Couteaux, R., 81, *107*

Daniels, B.A., 10, *25*
Davies, R.W., II, 11, *25*, *26*
Davila, H.V., *78*
Dawson, W.R., 28, *33*
DeBellis, A.M., *167*
Dedwyler, R.D., II, 20, *25*, *146*
del Castillo, J., 102, *107*
Delcomyn, F., 113, *144*
DeRiemer, S.A., 83, *107*, *109*

Derganc, M., 39, *49*
Derosa, Y.S., 37, *49*
Deschénes, M., 94, *107*
Dewhurst, S.A., *168*
Dodge, F.A., 98, *108*
Döll, W., *49*
Dowling, J.E., 82, 85, *107*, *108*
Dreyer, F., 81, *108*
Dudel, J., 102, *108*, 223, *223*

Eckert, R., 95, *108*
Ehinger, B., 150, 151, 152, *168*
Elliot, E.J., 213, 214, *225*
Emson, P.C., *169*
Erwin, V.G., *171*
Evans, P.D., 156, *168*, *169*, *171*

Falck, B., 150, *168*
Fambrough, D.M., 162, *168*
Fawcett, D.W., 43, *49*, 81, *107*, 149, 162, *168*
Fett, M.J., 232, *234*
Fernández, J.H., 46, 90, *108*, 175, 177, 179, 181, 182, 203, *225*
Fernández, M.S.G. 203, *225*
Fitzgerald, S., 8, *25*
Flacke, W. 156, *168*
Fonnun, F., *169*
Frank, E., 89, 90, 91, 95, 102, *108*, 283, *286*
Frazer, B.M., 284, *286*
Frazier, W.T.E., 114, *144*
Freytag, K., *49*
Friesen, W.O., 20, *25*, *26*, 37, *49*, 73, *78*, 97, *108*, *111*, 114, 118, 135, 138, 139, 140, 141, *144*, *145*, 146, 283, *286*, *287*
Fuchs, P.A., 75, *77*, 160, 163, *168*, 218, 221, 222, 223, 224, *225*, 283, *286*
Führer, J., 156, *168*

Gainer, H., *145*
Gardner, D., 87, *108*
Gardner-Medwin, A.R., 91, 95, *108*
Gaskell, J.F., 147, 150, *168*
Geffen, L.B., 82, *108*

Name Index

Gerschenfeld, H.M., 87, *108*, 162, 163, *168*
Giller, E., 149, 159, 160, 161, *168*
Gillis, C.N., 159, *171*
Goldman, J.R., 12, *25*, 51, 59, 65, *77*
Goodman, C.S., 149, *168*
Gorman, A.L.F., 114, *144*
Graf, L., *25*
Granzow, B., *109*
Grassé, P.P., 35, *49*
Gratiolet, P., 41, *49*
Gray, J., 8, *25*
Gray, R., 115, 128, 129, *144*, *145*
Greenawalt, J.W., *171*
Grinvald, A., 52, *77*
Grossman, Y., *233*
Gruner, O.C., 28, *33*
Gruol, D.L., *170*
Guthrie, P.B., 218, *225*

Hagiwara, S., 95, *108*
Hall, Z.W., *111*, 160, 161, *169*, *171*, *287*
Hammond, D.L., 10, *26*
Hamp, R., 118, *143*
Hanley, M.R., 160, *169*
Harant, H., 35, *49*
Harmon, L.D., 117, *144*
Harnett, J.C., 29, 32, *33*
Harper, G., *195*
Harris-Warrick, R.M., *170*
Harvey, W., 39, *49*
Haycraft, J.B., 32, *33*, 39, *49*
Henderson, L., 163, 164
Herbert, E., *169*
Hermann, E., *144*
Herter, K., 8, *25*
Heunius, J., 29, *33*
Heuser, J.E., 81, *108*
Hildebrand, J.C., 149, 155, 159, 160, 161, 165, *169*
Hillarp, N.-A., *168*
Hockfield, S., *235*
Hochstein, S., *226*
Hökfeldt, T., 158, 160, *169*
Horn, G., 30, *33*
Hoy, R.R., 227, *234*

Huang, L.M., *170*
Hubbard, J.I., 102, *108*
Huber, J., 3, *33*
Hudspeth, A.J., *111*, *171*, *287*

Ironmonger, J.W., 9, *26*
Ito, T., 150, *169*

Jansen, J.K.S., 59, 61, 71, 77, *77*, 78, 84, 97, 105, *108*, 199, 201, 215, 217, *225*, *226*, 232, 234, **249**, 281, 282, *286*, *287*
Johansson, O., *169*
Johnson, J.R., 27, 30, *33*

Kandel, E.R., 87, *108*, *109*, 114, *144*, 167, 199, 225
Karten, H.J., *170*, *286*
Katz, B., 99, 102, *107*, *109*
Kehoe, J., 87, 94, *109*, 162, *169*
Kennedy, D., *234*
Kerkut, G.A., 150, 151, *169*, *170*, *171*
Keyser, K.T., *170*, *283*, *286*
Kim, S.Y., 189
Kleinhaus, A.L., 94, *109*, 285, *286*
Klemm, D.J., 11, *25*
Kling, U., 118, *144*
Knight, B.W., *108*
Koester, J., *144*
Kramer, A.P., 12, *25*, *26*, 51, 59, 65, *77*, 95, *109*, 181, 182, 183, *195*
Köhler, G. **235**, *247*
Kravitz, E.A., 154, *167*, *168*, *169*, *170*, *171*
Kretz, J.R., 37, 48, *49*, 73, *77*, *283*, *286*
Kreutzberg, G.W., *110*
Kristan, W.B., Jr., *26*, 38, *49*, 50, 57, 72, *77*, 78, 91, *109*, *110*, *111*, **113**, 114, 128, 134, 135, 139, 140, *144*, *145*, 146, 218, *225*, 275, 282, *286*, *287*, *288*
Kuffler, S.W., 42, 43, 45, *49*, 61, *77*, 81, 87, 90, 95, 102, *108*, *109*, 149, 156, 161, *169*
Kupferman, I., *144*, 154, *169*, *171*
Kuwada, J., 182
Kuzmima, L.V., 150, *169*

Name Index

Landia, D.M.D., *108*
Lankester, E.R., 41, *49*
Larsen, W.J., 89, *109*
Lawler, A.R., *26*
Leake, L.D., *109, 170, 282, 286*
Le Douarin, N.M., 148, *170*
Lent, C.M., 150, 152, 153, 155, *170, 282, 284, 285, 286*
LePont, F., *26, 195*
Lesher, S., *77*
Levitan, H., 87, *109*
Lewis, E.R., 117, *144*
Linton, L., *25, 26*
Lissman, H.W., *25, 144*
Livanov, N., 42, 48, *49*
Livett, B.G., 82, *108*
Livingstone, M.S., 154, 155, 164, *170*
Ljungdahl, Å., *169*
Loewenstein, W.R., 95, *110*
Louis, P.C.A., 30, 31, *33*
Lowe, P., 29, *33*
Lundberg, J.M., *169*

Macagno, E.R., 42, 45, 46, *49*, 83, 87, 105, *107, 109*, 237, *285, 286*
MacDonald, J.F., 163, *170*
MacIntosh, F.C., 156, 160, *167, 170*
Magni, F., 91, *109*, 283, *286*
Magnusson, S., *25*
Mann, K.H., 32, *34*, 35, 37, 38, *49*, 118, *145*, 154, *170*
Manton, S.M., 7, *25*
Markwardt, F., 32, *34*
Marsden, C.A., 150, 151, *170*
Marttila, I., *144*
Mason, A., 87, *109*, 152, 154, *170, 282, 286*
Massoulié, J., 161, *167, 170*
McAdoo, D.J., 150, 155, 156, *170, 171, 287*
McCaman, R.E., *170*
McCamen, M.W., 150, 152, 153, *168, 170*
McDougall, W., 117, *145*
McGirr, S.J., *145*
McKay, R., 69, *78*, **235,** 235, 247, 284, 285, *288*

McMahan, U.J., 57, 75, 77, 79, 81, 83, 84, 85, *109, 110*, 149, *170*, 281, 282, *286*
Meech, R.W., 114, *145*
Meinertzhagen, I.A., *107*
Merickel, M., 115, *145*
Miledi, R., 99, *109*
Miller, J.P., 233, *234*
Milstein, G., **235,** 247
Minz, B., *34*
Miyazaki, S., 83, *109*, 205, 225, 226
Moor, H., *108*
Moquin-Tandon, A., 30, *34*
Morita, H., 95, *108*
Muller, K.J., 57, 61, 71, 72, 75, 77, *78*, **79,** 79, 81, 83, 84, 85, 89, 90, 94, 95, 97, 102, 103, 105, 106, *107, 108, 109, 110, 111*, 149, 162, *170*, 182, **197,** 197, 199, 203, 204, 205, 206, 207, 209, 211, 212, 213, 214, 217, 225, 226, 229, 231, 232, 234, **249,** 281, 282, 283, *286*
Mulloney, B., 177, *145*
Myhrberg, H.E., *168*

Nakajima, Y., 87, *111*
Neuhoff, V., *170*
Nicholls, J.G., 40, 42, 45, *48, 49, 50*, 52, 53, 55, 56, 59, 60, 63, 65, 71, 72, 77, *78*, 83, 85, 92, 93, 94, 95, 97, 98, 99, 100, 101, 102, 103, 104, 105, *107, 108, 109, 110*, 125, *145*, 149, 161, 163, *168, 169, 171*, **197,** 198, 199, 205, 215, 218, 220, 225, 226, 231, 233, 234, **249,** 281, 282, *286, 287*
Nistri, A., *167*

Obara, S., *146*
Olexa, S.A., *25*
Ono, J., *286*
Orchard, I., 48, *50*, 285, *287*
Ort, C.A., *26*, 42, 49, *50*, 73, 77, *78*, 110, *111*, 128, 129, 131, 134, *145*, 272, 275, 277, 280, 281, *287*
Osborne, N.N., 152, *170*
O'Shea, M., *168*

Osler, W., 32, *34*
Overstreet, R.H., *26*

Pappas, G.D., 90, *110*
Paré, A., 29, *34*
Parnas, H., *226*
Parnas, I., 63, 94, 107, 207, *226*, **227**, *233*
Paupardin-Tritsch, D., 163, *168*
Payne, J.F., 28, *34*
Payton, B.W., **27**, **35**, 95, *110*
Pecot-Dechavassine, M., 81, *107*
Pellegrino, M., 91, *109*, 283, *286*
Pentreath, V.W., 155, *167*
Peper, K., *108*
Peracchia, C., 90, *110*
Perez, H.V.Z., 150, *170*
Perkel, D.H., 117, *145*
Perry, W.L.M., 156, *170*
Petersen, T.E., *25*
Peterson, E.L. 125, 126, 127, *145*
Pfenninger, K.H., *111*
Phillips, C.E., 73, *78*
Pitman, R.M., 221, *226*
Poll, J., 150, *170*
Poon, M., *26*, *108*, *111*, 137, 138, 141, *144*, *145*
Potter, D.D., 43, 45, *49*, 90, 95, *109*, 160, *170*
Prichard, J.W., 94, *109*, 285, *286*
Pumphrey, R.J., *25*, *144*
Purves, D., 79, 81, 83, 85, 92, 93, 94, 97, 102, *110*, 215, 231, *234*

Rakic, P., 85, *110*
Ramon y Cajal, S., 201, *226*
Ready, D.F., *77*, 149, 163, *168*, *171*, 218, 220, 225, *226*, *286*
Reese, T.S., *108*
Retzius, G., 43, 44, 47, *49*, 52, *78*
Rieske, E., 89, *110*
Rinvik, E., *108*, *286*
Rolleston, J.D., 30, *34*
Roos, B., *144*
Rosenbluth, J., 87, *110*
Rubin, E., 48, *50*
Rude, S., 65, *78*, 149, 150, 152, 161, *171*

Saavedra, J.M., *167*
Salzberg, B.M., 52, 53, *78*
Sandri, C., *108*
Sargent, P.B., 69, *78*, 87, 93, *110*, 149, 150, 155, 156, 157, 158, 160, 161, 162, 163, 164, 165, *171*, 205, 283, *287*
Sawada, M., 152, 153, *171*, 281, *287*
Sawyer, R.T., **7**, 8, 9, 10, 11, 12, 13, 14, 23, 24, *25*, *26*, 174, *195*, **249**, 250, *275*
Schleip, W., 179, *195*
Schmucker, J.L., 30, *34*
Schnaitman, C., 160, *171*
Schneider, F.H., 159, *171*
Schubert, P., *110*
Schultzberg, M., *169*
Schwartz, J.H., 149, 159, 160, *168*
Scott, S.A., 71, 72, *78*, 89, 90, 95, 97, 106, *110*, 111, 170, 211, 212, 213, 226, 229, 231, 232, 234, 281, *286*
Sedden, C.B., *169*
Selverston, A., 233, *234*
Shafer, M.R., 283, *287*
Shelley, R.H., 11, *26*
Simpson, G.V., *145*
Skau, K., *167*
Smith, T.G., Jr., 114, *145*, *170*
Sommer, A., 150, *170*
Spitzer, N.C., *168*
Spua, M.E., *226*
Staehelin, L.A., 89, *111*
Standen, N.B., 114, *145*
Stent, G.S., 19, *25*, *26*, 41, *49*, 50, 72, 77, *78*, 91, 97, 98, 104, 105, *108*, *110*, *111*, **113**, 114, 118, 119, 121, 123, 125, 126, 127, 128, 138, 139, *144*, *145*, 146, 175, *194*, *195*, 217, 226, 262, *275*, 281, 283, *286*, *287*
Stewart, R., *109*
Stewart, W.W., 90, *111*, 240, *247*, 271, *275*, 283, *287*
Strumwasser, F., 114, *145*, *146*
Stuart, A.E., 41, *50*, 57, 73, 76, 77, *78*, 85, 87, 95, *111*, 129, *146*, 150, 156, *171*, 231, 234, 266, *275*, 280, 281, 284, 285, *287*
Stuart, D.K., *26*, 150, 181, 182, *195*

Suicide, 32, *34*
Sunderland, A.J., *109*, *170*, 283, *286*, 287
Szekely, G., 117, 118, *144*, *146*
Szerb, J.C., *34*

Talamo, B.R., *168*, *169*, *171*
Tauc, L., 87, *109*
Taxt, T., *108*
Teillet, M.-A.M., 148, *170*
Thieme, G., *168*
Thomas, B.E., 90, *170*
Thomas, L., 32, *34*
Thomas, M.V., *144*
Thompson, W.J., 41, *50*, 98, 104, 105, *111*, 119, 121, 123, 125, 126, 127, *145*, *146*, 217, *226*, 262, *275*, 281, 283, *287*
Tisdale, A.D., 87, *111*
Torp, A., *168*
Townsel, J.G., *169*
Toyoda, J., *108*
Tulsi, R.S., 87, *111*
Tweedle, C.D., *226*

Van Essen, D., 40, *50*, 59, 61, 71, 77, *78*, 83, 90, 105, *111*, 207, 217, *226*, 232, *234*, 281, 282, *287*
Van Orden, L.S., III., *171*
Vialli, M., 150, *171*
Vigny, M., *167*, *170*
Vitet, L., 30, *34*
Von der Wense, T., 150, *171*
von Euler, C., *144*
von Uexküll, J., 129, *146*

Wachtel, H., *167*
Walker, R.J., 152, 153, 154, 156, *169*, *171*
Wallace, B.G., 59, *78*, 92, 98, 99, 100, 101, 104, 105, *110*, 125, *145*, **147**, 150, 156, 157, 159, 161, *168*, *169*, *171*, 182, 198, 199, 200, 201, 202, 203, 217, *226*, 229, *234*, 280
Watanabe, A., 114, *146*
Waziri, R., *144*
Webb, R.A., 48, *50*, 285, *287*
Weeks, J.C., 138, 140, 141, *146*, 282, 283, *287*, *288*
Weinreich, D., *168*, *170*
Weisblat, D.A., **173**, 175, 182, 187, 189, 191, 192, 193, *194*, *195*, 233, *234*
Weiss, K.R., 155, *172*
Werman, R., *226*
Whitman, C.O., 173, 177, 179, *195*
Wiermaa, M.J., *167*
Wilkialis, J., *26*
Wilkenson, J.M., *171*, *287*
Willard, A.W., 140, *146*, 155, 164, *172*, *288*
Wolfe, D.E., 45, *50*
Wood, M.R., 82, *111*
Woodruff, G.N., *171*
Wrona, F.J., 11, *25*, *26*

Yaksta-Sauerland, B.A., 87, *111*, 152, 154, *172*
Yanagisawa, H., 39, *50*
Yau, K.-W., 37, 46, 48, *50*, 51, 58, 59, 61, 62, 64, 71, *78*, 83, 105, *110*, *111*, *171*, 207, *226*, 237, *247*, 266, *275*, 281, 282, 285, *288*
Yeoh, T.S., 156, *168*
Yokoi, E., 39, *50*
Young, J.D., *195*, *234*
Young, J.O., 9, *26*
Young, S.R., 139, *146*

Zackson, S.L., 187, *195*, *234*
Zdravic, F., 39, *49*
Zeman, G.H., *167*
Zipser, B., 41, *50*, 69, 74, *78*, 96, *111*, 158, *172*, **235**, 235, 239, 240, *247*, 284, 285, *288*

Subject Index

Ablation, 193
Acetylcholine (ACh), 69, 182–183
  hyperpolarization and, 87, 88
  receptors and, 165–166
  synthesis of, 155–156, 159–162.
  See also Neurotransmitters
Acetylcholine esterase (AChE), 159–162
Actinobdella inequiannulata, 10
Anatomy, 7–11
Annulus erector (AE) cells
  ACh release and, 156–157
  changes in synaptic potential and, 215–218
  description of, 52
  facilitation and, 102-106
  innervation of muscle fibers and, 73–76
  killing of, 57, 231–233
  lab exercise, 274
  nonrectifying junctions and, 97

  neuroreceptors and, 162–163
  origin of neuronal branching patterns, 182
  sensory receptors and, 73
  synaptic potential and, 214–218
Arborization, 59–60, 65

Batracobdella picta, 10–11
Behavior
  ecology and, 10–13
  embryonic, 183
  evolution and, 10–11
  feeding habits, 9–10
  foraging, 14–22
  ontogeny and, 23–24
  phylogeny and, 23–24
  swimming, 13
  systematics, 8–10
Bloodletting, 27–32
Bromolyseric acid, 152–153

316 Subject Index

Ca++
  endogenous polarization rhythms and, 114–115
  synaptic transmission and, 97, 102
  transmitter release and, 92–95, 97
Ca++ channels, 114
Cell 1. See Dorsal inhibitor
Cell 2. See Ventral inhibitor
Cell 3. See Dorsal excitor
Cell 4. See Ventral excitor
Cell 5. See Dorsal excitor
Cell 7. See Dorsal excitor
Cell 8. See Ventral excitor
Cell 27. See Oscillatory interneurons
Cell 28. See Oscillatory interneurons
Cell 33. See Oscillatory interneurons
Cell 102. See Dorsal inhibitor
Cell 107. See Dorsal excitor
Cell 108. See Ventral excitor
Cell 109. See Dorsoventral excitor
Cell 119. See Ventral inhibitor
Cell 123. See Oscillatory interneurons
Cell 204, 140–141
Cell 208, 140–141
Cell lineage, 187–193
Cell rhythm generation, 113, 114, 125–127, 135, 136
Chemical synapses
  between particular neurons, 95–100
  distribution of, 81–82
  physiology of, 91–95
  lab exercise on, 273
  structure of, 79–82
Chlorobutanol, 229–231
Choline acetylase, 159–160
Circuit analysis, 231
Cleavage, 177–179
Counterirritation, 29–30
Curare, 156
Cyproheptadine, 152–153

Diaminobenzidine (DAB), 84
Depression, 102–106
Development
  of leech, 173–181
  of nervous system, 181–193. See also Embryogenesis, neurogenesis
Dopamine
  hyperpolarization of N cells and, 69–72
  neuromodulation and, 154
  synthesis of, 155, 159, 182–183. See also Neurotransmitters
Dorsal excitor neurons, 131–134
Dorsal inhibitor neurons, 131–134
Dorsoventral excitor neurons, 131–134

Echothiophate, 161
Ecology, 10–12
Ectoderm, 177
Ectodermal precursor cell, 194
Ectoteloblast precursor cell, 189–191
Electrical synapses, 79–81
  between particular neurons, 95–100
  physiology of, 91–107
  regeneration of, 206–214
  structure of, 85–91
Electron microscopy, 85–89, 163, 203–205, 243–245
Embryogenesis, 175–180
  behavior and, 183
  definition of stages of, 175–177
Embryology, 173–175
Endogenous polarization rhythms, 114–115, 127
Evolution, 10–11
Excitatory postsynaptic potentials (EPSPs), 97
  facilitation and, 97, 98
  measurement of, 273
  N cells and, 215–217

Facilitation, 100–107
Faivre's nerve, 42, 277
  monoclonal antibodies against, 241–245
Fast Green, 228–231

Subject Index

Formaldehyde-induced fluorescence, 87, 150, 160
γ-Aminobutyric acid (GABA)
  depolarization of N cells and, 69
  neuromodulation and, 154
  synthesis of, 155–159.
  See also Neurotransmitters
Ganglia
  body-wall preparation, 264
  cellular organization of, 42–45, 52
  description of, 46–48
  monoclonal antibody staining of, 235–245
Gap junctions, 89–91
Germinal bands, 179–181, 187–188
Germinal plate, 177–179
Glial cell, 45
  gap junctions between, 89–91
  regeneration of, 213–214
Glutamate, 158–159
Gut formation, 179–181

Haementeria ghilianii, 10–13, 51
  cell lineage studies in, 193
  development of, 173–175
  foraging behavior, 20–22
  growth and feeding, 13
  identification of N, T, and P cells, 52–57
  neural differentiation and, 181–182
  neural branching patterns in, 57–59
  receptive field boundaries, 59–61
  swimming behavior of, 11–13
Haemopis grandis, 10–13
Haemopis marmorata, 10–13
Haemopis terrestris, 10–13
Head. See Subesophageal ganglion
Heart, 118–121
Heartbeat, 113–127
Heart excitor (HE) cells, 74–75, 98, 119–125
  HN cells and, 121–127
  inhibitory input and, 121, 122
  laboratory exercise on, 263
  neurotransmitter synthesis and, 155–159
  phase progression of activity and, 122–125
Heart interneurons (HN), 98–99, 121–125
  cycles of, 125–127
  facilitation in, 102–106
Helobdella stagnalis, 10–11
Helobdella triserialis, 10–11, 173–183
Hemiolepsis marginata, 10–11
Hirudin, 32–33, 39–41
Hirudo medicinalis
  atlas of neurons of, 277–288
  development of, 179–181
  heart of, 118–121
  leeching and, 31
  morphology of, 35–41
  photosensory organ of, 72–73
  receptive fields and, 65–69
  semi-intact preparation of, 129–131
  subesophageal ganglion of, 60–63
  swimming studies and, 135–136
Hoechst stain, 187
Hoover (HO) cells, 65–69
Horseradish peroxidase (HRP)
  electrical synapse localization and, 90–91, 106–107
  laboratory protocol for staining with, 267–271
  monoclonal antibodies and, 239–241
  neuron shape studies and, 57–58
  oscillatory interneuron detection and, 136
  peripheral terminal identification and, 153
  regeneration of synapses and, 207
  sensory terminal studies and, 63–65
  staining of ganglia and, 83–87
Humoral theory of disease, 29–30
5-Hydroxytryptamine (5-HT)
  depolarization of N cells and, 69
  iontophoresis and, 162–163

in R cells, 150–156
swim initiation and, 140
synthesis of, 155–156, 182
5-Hydroxytryptophan decarboxylase, 162–166

Immunological identification. *See* Monoclonal antibodies
Inhibitory postsynaptic potentials (IPSPs), 98
  measurements of, 263
  regenerating neurons and, 201
Innervation, 59–60, 232–233
Interneurons, 125–127, 135–136, 140–141

K^+ channel, 114
K^+ conductance, 72
Killing cells, 227–229

Laboratory exercises, 249–275
Leeching, 27–32
Leydig cells, 45
Locomotion, 7–9, 11–13, 52–57
Longitudinal (L) cells
  branch point failure in, 100
  changes in synaptic potential and, 215–219
  circuit analysis and, 231
  connections between cells in culture, 221–222
  facilitation and, 102–106
  innervation of, 85
  laboratory exercise with, 273
  neuroreceptors and, 162–164
  nonrectifying junctions and, 91
  rectifying junctions and, 91
  regeneration of, 199–201
  synaptic ultrastructure of, 87–89
Lucifer Yellow, 90, 233, 241, 271

Macrobdella decora, 9–10, 32
Macromeres, 178–181, 190–191
Magnesium ion. *See* Mg^{++}
Marsupiobdella africana, 10–11
Mechanosensory cells, 52–72.

See also T cells, P cells, and N cells
Merocyanin, 52–53
Mesoderm, 176–179
Mg^{++}
  neuromodulation and, 97–98, 154
  synaptic potentials and, 215–218
  transmitter release and, 92–93
Micromeres, 178–179, 189–191
Monoclonal antibodies
  HRP staining and, 238–241
  Lucifer Yellow staining and, 238–241
  N cell staining and, 235–245
  P cell staining and, 235–245
  supraesophageal staining and, 235–245
Monosynaptic connections, 91–94
Motor neurons
  branching patterns of, 73–75
  muscle fiber innervation and, 52–57, 75–77
  oscillator neurons and, 136–138
  swimming and, 131–134
Muscles, 52–57, 75–77

Na^+-K^+ pump, 71
Nerve roots, 42
Network rhythms, 115–118
Neurogenesis
  abnormal development and, 193
  behavior and, 188
  branching pattern origin and, 182–183
  cell lineage and, 187–193
  description of, 179–181
  differentiation and, 181–183
Neuromodulators, 154, 163–164
Neuropil glia, 192–193
Neurotransmitters
  chemistry of, 147–155
  synthesis of, 156–159, 182–183
Neutral Red dye, 87, 152
Nociceptive (N) cells
  changes in synaptic potential and, 214–218
  circuit analysis and, 231

connections between cultured cells and, 218–225
depolarizing agents and, 69
facilitation in, 98–106
K+ conductance and, 71–72
killing of, 228–231
laboratory exercise with, 261, 273
monoclonal antibodies against, 235–245
morphological features of, 57–59
neuroreceptors and, 162–165
receptive fields and, 65–69
regeneration and, 199–205
staining of, 83–85
terminals and, 65–69

Octopamine, 154–155
Oscillatory interneurons, 135–141
Oscillatory network of swim cycle, 136–141

Penile erector (PE) cells
electrical synapses and, 96
monoclonal antibodies against, 235–245
Peristaltic heartbeat mode, 119, 122–125
Phase angles, 117–118, 122–125, 133–136
Photosensory organ, 73
Placobdella ornata, 10–11
Placobdella parasitica, 24
Postsynaptic potentials (PSP)
electrical synapse identification and, 93
electrical transition and, 102
Posttetanic potentiation, 102
Potassium ion K+. *See* K+
Pressure (P) cells
branch point failure and, 102–106
changes in synaptic potential and, 214–218
connection between cultured cells and, 218–219
facilitation in, 102–106
K+ conductance and, 67–72
killing of, 228–231
laboratory exercises on, 261, 263
monoclonal antibodies against, 235–245
morphological features of, 57–59
neuroreceptors of, 162
posttetanic potentiation in, 102
receptive fields and, 63–65
rectifying junctions and, 91, 95–97
regeneration of, 199–201
stimuli for firing, 52–55
synaptic ultrastructure of, 85–89
terminals and, 65
Presynaptic inhibition, 97–98
Presynaptic modulation, 122–125
Procion Yellow, 85
Pronase, 57
Protease
ablation studies and, 193
deleting single cells and, 231–233
injection technique with, 227–228
reinnervation studies and, 62–63

Receptive field organization, 54–61
Reciprocal inhibition networks, 117
Rectifying junctions, 91
Recurrent cyclic inhibition networks, 117–141
Regeneration, 61–63, 197–225, 227–228
Reinnervation, 61–63, 77
Retzius (R) cells
connections between cells in culture, 218–224
development of, 152–153
electrical coupling of, 258, 263, 273
electrical synapses and, 95
functional role of, 152–155
5-HT and, 150–152
injection of protease into, 228–231
neurotransmitter synthesis in, 155–162
nonrectifying junctions and, 91
swimming and, 140–141
synaptic ultrastructure of, 87–89

187, 193
-10
nents, 113–143. See
  entral rhythm gener-
  swimming

cell macrophages, 203
.tation, 37–38, 179–181
.citation networks, 115
-intact preparation, 129–131
.sory (S) cells
  branch point failure and, 105
  circuit analysis and, 231
  electrical synapses and, 95
  laboratory exercise and, 261
  Procion Yellow and, 89
  rectifying junctions and, 91
  signaling by, 69–72
Serotonin. See 5-Hydroxytryptamine
Stem cells, 177–179
Stretch receptors, 72–73. See also Hoover cells
Subesophageal ganglion, 46–48, 61
  formation of, 179–181
Supraesophageal ganglion
  cell lineage of, 188–191
  formation of, 179–181
  monoclonal antibodies against, 235–245
Swimming
  as a rhythmic movement, 113, 128–129
  central rhythm generator of, 135–136
  initiation of, 128–129, 135–136, 139–141
  motor neuron activity and, 73–75, 131–134
  oscillator network and, 136–139
Synapses. See Chemical synapses; Electrical synapses
Synaptic formation, 182
Synaptic vesicles, 87
Synchronic heartbeat mode, 118–121, 122–125
Systematics, 7–10

Tail ganglion, 48
Tetraethylammonium (TEA$^+$), 94, 274
Tetraisopropylpyrophosphoramide (ISO-OMPA), 161–162
Teloblasts, 177–179, 188, 191–192
Teloplasm, 177–179, 194
Theromyzon rude, 10–11, 175
Touch (T) cells
  branch point failure in, 105–107
  changes in synaptic potential and, 214–218
  circuit analysis and, 231–233
  connections between cultured cells and, 160–162
  electrical synapses between, 89–91
  facilitation in, 98–102
  K^+ conductance and, 72
  killing of, 228–231
  laboratory exercises with, 261, 263, 265
  morphological features of, 57–60
  neuroreceptors and, 162–166
  posttetanic potentiation in, 102
  receptive fields of, 65
  rectifying junctions and, 91, 97–98
  regeneration of, 199–205
  sensory terminal morphology of, 63–65
  staining of, 83–85
  stimuli for firing, 52–57
  swimming and, 139–140
Transmitter release, 90–106

Use of leeches
  contemporary, 32–33
  historical, 27–33

Varicosities, 63, 83, 154
Venesection, 30
Ventral blood sinus, 42
Ventral excitors, 131–134
Ventral inhibitors, 131–134

Wet cupping, 30